PERMACULTURE

Practical & Radical Solutions for a World in Crisis

Mira Illeris and Esben Schultz

Permanent Publications

Published by
Permanent Publications
Hyden House Ltd
13 Clovelly Road
Portsmouth
PO4 8DL
United Kingdom
Tel: 01730 776 582
Email: enquiries@permaculture.co.uk
Web: www.permanentpublications.co.uk

Distributed in North America by Chelsea Green Publishing Company,
PO Box 4529,
White River Junction,
VT 05001, USA
www.chelseagreen.com

Distributed in Australia by Peribo Pty Limited,
58 Beaumont Road,
Mt Kuring-Gai,
NSW 2080, Australia
https://peribo.com.au

Cover artwork and book design by Two Plus George Limited,
info@twoplusgeorge.co.uk

Printed in the UK by Bell and Bain, Thornliebank, Glasgow

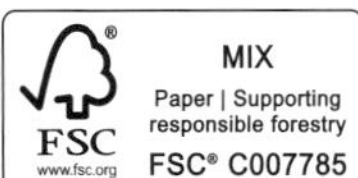

This product is made of material from well-managed FSC®-certified forests and from recycled materials and other controlled sources.

The Forest Stewardship Council ® (FSC) is a non-profit international organisation established to promote the responsible management of the world's forests. Products carrying the FSC label are independently certified to assure consumers that they come from forests that are managed to meet the social, economic and ecological needs of present and future generations.

British Library Cataloguing-in-Publication Data
A catalogue record for this book is available from the British Library

ISBN 978 1 85623 258 6

Praise for the book

Any thoughtful farmer or gardener who knows their soil, and observes nature will inevitably find themselves questioning how they farm. Their thoughts will move to the complexity, abundance and stability of forests; and then to imitating and learning from them; and then to the ideas of permaculture. Turning those ideas into a system which is humanly, economically and sociably sustainable becomes an all consuming challenge for a growing number of thoughtful farmers. This comprehensive, pragmatic and practical book makes a wonderful guide to the start you on that journey.

Guy Singh-Watson,
Founder of Riverford

A compelling and deeply thoughtful exploration of how permaculture can respond to the ecological and social challenges of our time. Blending big-picture vision with grounded, practical insight, the authors guide readers through the principles, systems, and real-world applications that make permaculture such a powerful framework for regeneration. What makes this work particularly valuable is its ability to connect the practical realities of farming and land stewardship with the wider societal shifts needed to create a resilient future. This is an inspiring and important contribution for anyone interested in regenerative agriculture, permaculture design, or the practical pathways toward a more balanced and abundant world.

Rachel Phillips,
Director of The Apricot Centre

In a world starved of good food from good soil, this book provides the recipe for small-scale solutions that can proliferate widely and form the foundations for a new decentralised food system.

Pete Russell,
Founder of Ooooby and
Cofounder of DecentAlliance.com

Permaculture, at its heart, is about intelligent design: learning from nature's billions of years of research and development to create systems that are not just sustainable, but regenerative. Imagine a world where businesses are structured like thriving ecosystems, where waste is a resource, and collaboration is inherent. Envision cities that manage water with the wisdom of a forest, and energy systems that harness the sun's generosity with elegant efficiency. This book serves as an important invitation to explore these possibilities. It reminds us that the power to cultivate a more resilient and harmonious world lies within our grasp, requiring not just new tools, but a fundamental shift in our collective mindset. It is another call to action for all sectors of society to embrace the profound intelligence of nature and to design our lives, our businesses, and our communities accordingly. The time for this regenerative transition is not just now; it is overdue.

Perrine Bulgheroni,
Cofounder of Bec Hellouin

Mira and Esben take a grand tour through the ideas, ethics, and techniques that form 'The Great Permaculture Vision', encompassing water, energy, construction, organisation, monetary systems and most of all land cultivation. I love their down-to-earth style that does not shy away from taking a clear stand on contentious topics like ruminants and no-till and tree-based agriculture. This will be an inspiration for those searching for positive ways forward in our chaotic world.

Martin Crawford,
Founder of Agroforestry Research Trust

A clear and inspiring introduction to permaculture in a Nordic context, where practical knowledge meets ecological philosophy. It invites a reimagining of how we live by rooting food production close to where people are. Through layered food forests, photosynthesis is optimized and carbon sequestration is increased, while enabling greater food production per hectare and potentially freeing land for nature and the restoration of ecosystems. Ultimately, the authors point towards a more regenerative and interconnected existential paradigm.

Isabelle Denaro,
Filmmaker, Director, Documentarist

Thank you

To all of you who are working to establish permaculture projects all over the world and are thus at the forefront of the transition to a future that regenerates natural resources and counteracts climate change.

To all of you who have contributed photos from many, many permaculture projects and have thus helped to ensure that the book now appears as a dynamic celebration of the rapid spread of the permaculture movement. There are so many of you that we cannot mention all of you by name and project here, but you are mentioned along the way in the book.

To Tony Andersen and Poul Erik Pedersen for the knowledge you have given us on the Permaculture Design Course.

To Permanent Publications for cooperation on this publication, feedback and designing this English version.

For proofreading and constructive feedback on specific subject areas: Andreas Kamp PhD; Freja Nygaard Rasmussen, engineer in life cycle calculations for construction; Bjarne Grube Wickstrøm, pioneer in natural building; Rasmus Hougaard Nielsen, economist contributor in Positive Money; Mathis Kærn Berggreen, also active in Positive Money; Jacob Rask, economist; Petter D. Jenssen, professor of wastewater management and compost toilets; Frederica Miller, permaculture architect; Alexandra Rempel, MArch, PhD, associate professor; and Mette Vaarst, senior researcher at Department of Animal and Veterinary Sciences.

To illustrator Sune Watts for illustrating our interpretation of the permaculture principles. For general feedback, Birgitte Simonsen and Knud Illeris.

To the members of Skovvirke for creating our permaculture farms; to Svanholm for making Skovvirke possible and for being good neighbors thereafter.

Contents

About the Authors

Mira Illeris and Esben Schultz live in Denmark at Reforest Farm, where they work with food forests, chicken forests and nut forests, and where they have built their off-grid straw-bale house and hold courses. The farm is located in a cluster of six cooperating permaculture farms, 'Skovvirke', which they started in 2011 and received planning permission for in 2015.

Before this, they were the initiators of The Self-Sustaining Village on Funen, which started in 2004, where they have worked with permaculture design for the farming systems and designed and built their own straw-bale house. In 2008 they were awarded the permaculture diploma for this work.

In 2009 they started the Scandinavian Permaculture Magazine, which they published for 13 years before it was transferred to the national permaculture associations.

The authors are both trained organic farmers and have roots in agriculture. Esben Schultz grew up on a family farm in Jutland, while Mira Illeris grew up in the organic farming collective Svanholm on Zealand.

Back in 2001 they took the permaculture design certificate (PDC). At that time, the spread of practical permaculture projects was very limited in Denmark, and they went on a study trip to England, where, among others, Martin Crawford's forest garden, Tinkers Bubble, Ragmans Lane Farm and The Anchorage were a great inspiration for their further work. Since then, they have visited many inspiring permaculture projects in Scandinavia and across Europe.

Today they are active in the Danish Permaculture Association where, together with others, they work for political influence to spread and improve the conditions for permaculture farming. They have regularly given lectures, participated in debates and written columns and debate articles about permaculture.

Editions of this book were first published in Danish, Swedish and Norwegian in 2019 and were expanded and revised in 2023.

About this Book

The book is divided into eight chapters.

In the first chapter, 'What Is Permaculture?', you will find a clarification of the concept, the permaculture ethics, the principles of permaculture, the design process and a brief introduction to how the four central natural resources have been degraded and the organization that has led to this – called the five catastrophes. This section provides, in a concentrated form, both the theoretical and practical starting point for the concept of permaculture.

The next six chapters deal with the topics 'Cultivation', 'Water', 'Energy', 'Construction', 'Organization' and 'Monetary Systems'. Each of these topics starts with a brief outline of the permaculture vision for the topic, which clarifies the overall direction we are going in. This is followed by theoretically based knowledge, which is consistently linked to concrete solutions that we can use in our permaculture systems – and which the reader can be inspired by and thus actually try to put into practice. In the final chapter, 'The Great Permaculture Vision and the World Today', we conclude by putting the transformative work of the permaculture movement into perspective in relation to system-preserving mainstream politics.

To make the book easy to understand, the main sections are divided into many small subsections with clear headings.

This book is based on the climatic conditions in the temperate world and is aimed at anyone who wants to become acquainted with permaculture and get an overview of the concept based on current research. At the same time, it presents the latest techniques and solutions and will thus provide new knowledge, probably even to the experienced permaculturist.

Introduction

Permaculture: a real alternative

Permaculture is a holistic approach that provides the techniques and tools needed to regenerate natural resources and counteract climate change, which is already well on its way to destroying our natural foundation and throwing the world into an unmanageable and catastrophic situation.

Permaculture provides the necessary countermeasure by working *with* nature instead of *against* it. With permaculture, we can achieve not only a carbon neutral society in a relatively short period of time, but even a society that directly counteracts climate change by storing carbon.

In this book, we will first and foremost provide an easily understandable and up-to-date introduction to the many aspects of permaculture; to what permaculture is and to how you and everyone else can participate in the permacultural transition.

This book builds on the previous work of a large number of permaculturists. This applies to Bill Mollison and David Holmgren, who developed the concept. Since then, there has been development on how permaculture principles can be applied in practice in temperate climates. Here, we would particularly like to mention Patrick Whitefield's book, *The Earth Care Manual,* and Maddy and Tim Harland and the Permanent Publications team for their many years of publishing *Permaculture* magazine, which has consistently shared the movement's growing knowledge for over three decades.

The contemporary and global situation that permaculture must relate to has changed dramatically in the past two decades, however, and in particular, the seriousness and extent of climate change have become clear and omnipresent. We therefore believe that there is a need for a book that simultaneously provides an overview of the concept of permaculture and, based on current research, consistently addresses climate change – and how we can counteract it in practice.

We live in a time when most agricultural land is already degraded and much of our water and air is already polluted. We live in a time when the amount of greenhouse gases in the atmosphere is already too high and climate change is affecting the world in a way that could end up making it uninhabitable for humans.

Therefore, we need to get to a point where everything we do contributes to regenerating natural resources and storing carbon. Today, through most of our everyday actions, we contribute to worsening the climate situation, and we need to reverse this, so that when we grow our food and wood for timber, when we build our houses, when we produce our electricity, etc., we must do so in a way that stores rather than emits carbon. We need to create an economy and develop an organization that enables the transition to a carbon-storing lifestyle and a carbon-storing society.

Converting our agricultural systems to food forests and agroforestry holds the greatest potential for carbon storage. When we do not cultivate the soil, but instead let trees and other perennial plants grow on our fields, CO_2 will be absorbed from the air and the carbon from this will be stored in the trees and in the soil as dead plant material, etc. Therefore farmers and small farms are particularly central to the transition. But permaculturists in the city are certainly also key, because it is only when they offer financial support by purchasing goods from permaculture farms or actively participate in creating rural-urban links, establishing food cooperatives or community supported agriculture, and fighting politically for fair conditions for permaculture farming, that the transition becomes practically possible for farmers.

Permaculturists in the city must demand that their food is produced in a carbon-storing way and help organize the circuit of food and nutrients. In this way, permaculturists outside farming play an active and significant role in ensuring that carbon storage takes place.

And finally, there is potential in food production in the cities, because a degree of self-sufficiency can be achieved here by using intensive permaculture cultivation techniques in private and community gardens.

The permaculture ethic is about fair sharing and solidarity with people in poor countries and with future generations; about solidarity with our children and unborn children who will inherit the Earth and who cannot fight for their rights themselves while the opportunity to tackle climate change is still there.

It is we, who are alive now, who must fight for their rights. We have the knowledge of what to do and what not to do to make the transition – and knowledge creates obligations. We want to be able to look our children, grandchildren and great-grandchildren in the eye and say we did everything we could. And we want it to succeed.

The permaculture movement started small and has now, over more than 40 years, developed an understanding and refined tools and methods to regenerate natural resources and counteract climate change. More and more people are participating in ways determined by their circumstances and the opportunities they have.

This book is based on the content of the Permaculture Design Course we took back in 2001, which was developed by the Australian Bill Mollison and taught by, among others, Tony Andersen, who was central to the founding of Permaculture Denmark. We have subsequently taught the course ourselves many times, each time developing it further and constantly striving to update it by injecting the latest knowledge, techniques and solutions.

Tony Andersen presented permaculture as a holistic approach to the regeneration of natural resources and claimed that, by following the permaculture approach, we would inevitably create such regeneration in the long term. At the time, it seemed very far-reaching to us that we should actually be able to absorb more CO_2 than we emitted.

But a few years later, the message from the renowned American climate scientist James Hansen was that it was no longer enough to 'just' reduce or stop the use of fossil fuels. To avoid self-reinforcing climate change, the concentration of greenhouse gases in the atmosphere had to be reduced.

It was this unequivocal message that made it clear to us that permaculture is not an unrealistic dream, but rather an unequivocal necessity.

Because permaculture works with nature's own ecosystems as a model, it provides long-term solutions to a wide range of problems that global society is struggling with: climate and biodiversity crisis, groundwater pollution, soil erosion, food insecurity, etc.

Therefore, we strongly recommend that permaculture methods should be spread over large areas and designs and implementation plans must be made for the conversion of entire bioregions and countries to permaculture. This book, based on calculations for diet, yield figures, carbon storage, etc., provides a progressive case for how permaculture can feed us and reverse climate change.

This example helps to emphasize that if we as a global society choose to convert our agricultural systems to permaculture, we will actually be able to tackle the biodiversity and climate crisis, which for many people may otherwise seem insurmountable or even inevitable disasters. The first step to change is to be able to see the vision before us.

As we engage in tackling the enormous challenges of our time in relation to climate, environment and social conditions, it is our experience that the overall thinking and general principles of permaculture can help us to target our work more efficiently and solve problems more effectively.

Permaculture shows us how we can have a society and a way of life where the problem of climate change is solved. By engaging in permaculture, we constructively enter into this historically gigantic regeneration task, and that is why the new generation of young permaculturists has also been called generation regeneration.

The possibility to regenerate natural resources and counteract climate change means that it is not too late. Nature has been thrown out of balance, but if we take systematic action, we can restore this balance.

Thus, permaculture also depends on a new view of humanity's interaction with nature. History shows that where humans have settled and created agricultural systems, we have left behind deserts and degraded nature, which are often now conservation areas, being protected from further human damage. But now we as humans and permaculturists must go out onto the degraded agricultural lands and regenerate them.

We must reforest the Earth. And legislation must therefore be changed to reflect the fact that humans are a necessary actor in the degraded agricultural landscape, so that we as the permaculture movement can jointly establish forest gardens, food forests and forestry.

If the climate challenge is to be solved globally, rich countries must take the lead, and if the rich countries are to take the lead, there must be citizens in those rich countries who take the lead and show politicians that there is a demand for change. It is much more powerful to not just say that someone should do something about the problem but to also show through our actions that we are doing something about it ourselves.

Permaculture is now ready with a toolbox that gives us the tools to do what is needed. Permaculture is a real and attractive alternative to overconsumption, pollution, loss of biodiversity and climate change. And the transition is already underway.

A house integrated with a vegetable garden and chickens at the permaculture center Växhuset near Söderhamn, Sweden.

RALF PALMPERS

1

What is Permaculture?

Concept clarification

What is special about permaculture in relation to the more mainstream concepts of 'ecology' and 'sustainability'? And what is specific to the permaculture movement when compared to the more broadly defined environmental movement?

Ecology

The term ecology originates from the Greek words *oikos*, meaning 'household', and *logos*, which can be translated as 'study of'. Loosely translated then, ecology is the study of the household of nature. Ecology is thus the science of the relationship between living beings, and between living beings and their environment.

Sustainability

The concept of sustainability was popularized by the Brundtland Report, published in 1987. Sustainability entails that meeting the needs of current generations cannot come at the expense of the ability of future generations to fulfill their needs. So, when something is classed as sustainable, it is implied that there is no deterioration of natural resources, including land, water, air and energy sources.

Permaculture

The term permaculture originated in 1978 when Australians Bill Mollison and David Holmgren published the book *Permaculture One*. Permaculture is based on the premise that it is not sufficient to preserve our resource base as it is. The degradation of the natural environment is now so advanced in terms of soil erosion, lack of clean water, air pollution, biodiversity loss and climate change, that what we face is a task of rebuilding, also known as regeneration.

A central element of permaculture is establishing new lush plant systems because these can regenerate topsoil; increase biodiversity; clean the soil, air and water; build up more energy in the form of biomass than is used and counteract climate change by storing more carbon than is emitted in the form of CO_2.

Organic farming is based on traditional farming methods which, amongst other things, include plowing the soil and cultivating annual crops. Requirements are placed on organic farms to operate without the use of pesticides and artificial fertilizers and to have better animal welfare, among other things.

Permaculture farming and gardening take a completely different approach. Nature's own ecosystems are always sustainable, and it is these that permaculture uses as models. In permaculture, however, we manipulate the model in order to make permaculture systems more productive for human needs. This is achieved through intentional design, which could be, for example, the design of food-producing forests, also known as agroforestry, food forests and forest gardens.

Furthermore, permaculture uses nature and its cycles more broadly as a model for designing agricultural systems, settlements, buildings, rural-urban links, monetary systems, and more, which meets the needs of humans, and the Earth.

Resilience

Resilience is another key concept. Permaculture systems are designed to be resilient – like natural ecosystems – to external stresses such as drought, heavy rainfall, disease, lack of fossil-fuel supply, the loss of income from a single source and so on. In a time of advanced climate change, rising energy prices or failing supply chains and the resulting political tensions and wars, the need for resilience appears far more present and important than in previous decades.

By designing diversified local cycles, we become independent of outside inputs. Several permaculture principles underpin this resilience.

Permaculture ethics

All cultures have ethics, but the ethics are not always explicit or even conscious. Ethics lead to norms of what is right and wrong. These norms guide our choices, which ultimately lead to our concrete actions and practices. Permaculture is guided by a three-part ethic, which gives direction and intention to a permaculture project:

Earth care
People care
Fair share

We need to remember all three aspects of this ethos. For example, if we craft a beautiful farming project that regenerates natural resources and thus cares for the Earth, it is important that the people working on the project aren't overburdened or worn out in the process. We must aim to create structures that allow the people in permaculture projects to thrive.

The project must also take care of us humans by covering our needs for food, water, medicine, timber etc. For example, if we have several hectares of arable land that are cultivated extensively and therefore give a low food yield, it is – in a world as densely populated as the world is today – not caring for people in poorer nations. They can't afford high food prices, and this slips into the ethics of fair share. Our permaculture systems need to be both highly productive and labor-saving in order to care for people and create equity. Highly productive cultivation systems will then leave space for wild nature in other places.

Conversely, a social or pedagogical project which cares for people but, for example, receives food, goods and labor from all over the world via air transport, cannot be said to be taking care of the Earth.

In practice, fair share in wealthy parts of the world means a reduction in the consumption of goods, travel, etc., while in developing countries, it could lead to an increase in consumption. People in wealthy countries need to redistribute their surplus of knowledge and material resources.

The climate crisis makes it clearer than ever that the current distribution in the world is not fair. There is a clear link between wealth and climate impact at a global level, with the top 10% of income-earning individuals accounting for 50% of humanity's consumption-based emissions, and the bottom half of the world's population accounting for 8% of humanity's consumption-based emissions.[1]

HELEN ILES livinginthefuture.org

In Wales, a special planning policy has been made for permaculture projects, 'One Planet Development', which makes it possible to build and establish permaculture farms in the rural zone when a number of conditions are met.

You must be able to cover your minimum needs for income, food and energy through agricultural production or other use of the site's resources. The houses must also be built from the site's resources and fit into the landscape.

Finally, you commit to calculating your ecological footprint and not using more than your fair share of the world's resources. The picture is from Lammes Ecovillage, whose initiators helped to get the policy made. Many permits for One Planet Developments have since been granted around Wales.

It is unfortunately also the case that the consequences of climate change trend towards having the worst impacts in many of the world's poorest nations, while the wealthy nations that have created and indeed exacerbate the climate crisis are not hit as hard. Following the ethic of fair share, the populations of the world's richest countries have an obligation to make especially large contributions towards climate amelioration, even though it means that their consumption will be reduced.

Permaculture has no metaphysical superstructure and it is not affiliated with any specific religious denomination or political ideology. Permaculture can therefore be used by people of all faiths and none.

The ecological footprint

– A tool for calculating Earth care and fair share

The permaculture ethic of *Earth care* must be understood holistically. It is about creating a lifestyle and societies that can regenerate all four natural resources: clean water, clean air, fertile topsoil and energy through carbon storage.

A useful calculation method in this context is the ecological footprint. The ecological footprint calculates the area of land and/or sea required to produce the resources that a population or an individual uses, and to absorb its waste.

The ecological footprint of the average citizen, worldwide, stands at 2.7 global hectares, and with only 1.5 global hectares currently available per citizen of Earth, there is an overconsumption of natural resources overall. We would need 1.75 Earths to cover humanity's consumption in a sustainable manner.[2] This overconsumption, along with global population growth, is resulting in a steady decrease in the number of global hectares available per citizen.

Overconsumption occurs primarily as a result of burning fossil fuels; that is to say, resources that were produced in the distant past. The consumption of renewable resources is also included. Even though biofuels or food can be produced completely without the use of fossil energy, and can therefore be said to be CO_2 neutral – because as much CO_2 is absorbed by the plants as is emitted when they are burned – there has still been a consumption of land during their production. And land is a limited resource that the world's population must share. More than half of the global ecological footprint is due to our use of fossil fuels, expressed in the model as the area of new forest needed to absorb the CO_2 emitted.

Of course, it is a simplification to include all environmental impacts in one number as we do when calculating an ecological footprint; it should therefore be supplemented by looking separately at all four natural resources – soil, energy, air and water – which permaculture projects aim to enhance. The consumption of groundwater, for example, is not included in the ecological footprint at all.

Clean water supplies are an acute problem in many countries, making it important to consider whether a permaculture project regenerates the resource of clean water. Soil contamination, air pollution and nuclear waste production are also serious environmental problems that are not included in an ecological footprint calculation.

If we remember to examine these other environmental issues separately, the ecological footprint is a great tool to help us get an overview of how far we have come towards creating a permaculture project or a permaculture lifestyle. If you are a citizen of a society with a high ecological footprint, it will require a great deal of focus to bring your ecological footprint down to an equivalent of the resources of just one planet. If you manage this, it will be fair to assume that you are not exceeding nature's limits in other areas either, even though you are not making calculations for them. An ecological footprint calculation shows us roughly how much effort is needed. It's a good tool to give us an overview of where it's important to take action and where it's less important.

You can calculate your personal ecological footprint on the website www.footprintcalculator.org.

Ecological footprint and economic inequality

The calculation of the ecological footprint, where we quantify environmental impact, has the advantage of making it much easier to compare just how much environmental impact and resource consumption different countries in the world are responsible for, and whether they are using more than is fair from a distributional perspective.

Just as the richest people have the largest climate footprint, they also have the largest ecological footprint. Globally, it is the low- and middle-income countries in South America, Central America and Africa, as well as India and Indonesia that have the smallest ecological footprints per capita. Countries with stronger economies, where there are more resources to initiate a transition away from fossil fuels, have an excessively large ecological footprint. For example, it would take 2.4 planets for everyone to live like the population of the UK, and almost five planets for everyone to live like the populations of the US, Canada and Denmark.[3] Here, resource consumption is so far from sustainable levels that we can only conclude that a complete transformation of these societies and the lifestyle of their populations is needed if they are to live sustainably and not take more than their fair share. It's not just a matter of small adjustments. The same trend applies within individual countries, where it is also the richest inhabitants who have the largest ecological footprint.

The five catastrophes

Permaculture is about creating holistic, long-lasting solutions that regenerate natural resources. It has become necessary to regenerate natural resources because we have degraded them over time and continue to degrade them today. Before we get into visions and solutions, it is therefore relevant to describe the conditions that have created what in permaculture are called the five catastrophes.

Each of the four natural resources, water, land, air and energy, are linked to one or more catastrophes that have degraded the resource. These catastrophes are highly interrelated, and climate change is linked to the degradation of soil, water, energy and air. The last catastrophe is the organization of society, which is to blame for all these catastrophes.

It is often portrayed that these catastrophes are an inevitable consequence of growing food for many, building good housing etc. but it is important to make it clear that the catastrophes could have been completely avoided and that the world's population could at the same time have lived good lives with high prosperity and far greater security than today. It is fundamentally about the approach, about working *with* nature instead of *against* it.

We will briefly review each catastrophe here.

The water catastrophe

Clean fresh water is a limited resource in many places in the world. Two billion people in the world lack access to safe drinking water, and about half of the world's population experiences severe water scarcity at some point during the year.[4]

For the world's poor, the lack of water contributes to inadequate sanitation, which leads to the spread of diseases that could otherwise be easily avoided, e.g. cholera.

Many of the water systems that are important for ecosystems to thrive and that supply a growing population have become stressed. Rivers, lakes, wetlands and aquifers dry up or become polluted.

Climate change is helping to worsen the situation in the form of altered weather patterns that cause droughts in some areas and floods in others, just as the increased temperatures reduce the extent of the glaciers that supply large parts of the world's population with meltwater.

50% of the world's population gets drinking water from groundwater.[5] Groundwater is formed when water that hits the Earth's surface as precipitation seeps down through the soil layers until it reaches the groundwater zone, where all cavities are filled with water. On its way down through the ground, the water is purified and absorbs salts and minerals, making it good to drink: deeper down the water is too salty, as it is closer to the surface by the coast.

The resources of clean groundwater are degraded by using more of it than is formed and by polluting it. When we pump up more water than is formed, it will result in the groundwater table falling. This often happens around the larger cities and where there is a large consumption of water for irrigation in agriculture. 70% of the groundwater used by mankind is used for irrigation.[5]

If we buy food and textiles imported from dry areas of the world where the products are grown on irrigated fields, we are pushing to exacerbate the problems of water shortages in these areas.

Overexploitation of the groundwater resource means that we have to go ever deeper, or further away from the city, to get enough water to cover our consumption, and this can also result in streams drying up, which is destructive to nature.

Contamination of groundwater occurs with chemicals from landfills and industry; PFAS (Per- and polyfluoroalkyl substances) from firefighting training facilities; from the sea; pesticide residues and nitrate leaching from agriculture; as well as by salt water penetrating the groundwater in coastal areas as a result of rising sea levels.

In the EU countries today, 29% of the total groundwater area cannot cover the ecosystem's or society's need for water, either because the quality of the groundwater has deteriorated or because the groundwater table has been lowered.[6]

The air catastrophe

99% of the world's population breathes air that does not meet the World Health Organization's (WHO) guidelines for healthy air. Air pollution is thus among the biggest environmental threats to human health. Air pollution damages our health by, among other things, creating a greater risk of respiratory and cardiovascular diseases. But air pollution affects the world's population disproportionately, as 89% of premature deaths due to air pollution occur in middle- and low-income countries.[7]

The air is polluted with particles and various gases from factories, power plants, wildfires, traffic and wood-burning stoves, for example. Poor combustion with wet firewood and low temperatures in just a single wood-burning stove can destroy air quality locally.

Trees help clean the air in their immediate surroundings. They produce oxygen and trap particles on the surface of their leaves, which are washed to the ground when it rains.[8] Nevertheless, it is of course best to avoid creating particle pollution in the first place.

The catastrophes associated with soil

If we look at the Middle East, where agriculture started approximately 12,000 years ago, it is surprising that this particular desert landscape was once the cradle of agriculture. Similarly, it is curious that the bare rocks of Greece were once a place of lushness and home to a magnificent culture. It may also seem strange that there were once farmers in the Arizona Desert with its characteristic cacti. The list of once great cultures that have left behind deserts today is long. But what is the story behind these sand and rock deserts?

The topsoil is created

The soil's constituents of stone, gravel, sand and clay – its texture – are created by weathering over long periods of time as glaciers have broken the bedrock into smaller parts and as the weather alternates between frost and thaw. From this texture, the plants have created topsoil with decomposed plant material, fungi and organic life of small organisms and larger animals such as earthworms. Plants – from small herbs to tall trees – generate the topsoil by depositing dead plant material and being a niche for fungi and microorganisms in the soil. At the same time, the plants constantly hold onto the topsoil.

Erosion and desertification

In agriculture, the soil is left bare after plowing. Without the plants to protect it from drying out, the impacts of rain, and without their roots to hold the soil's small particles, the soil will erode away. There will be no food for the fungi and organic life here. Wind will blow away the small clay and sand particles and rain will wash them away. The particles will end up in valleys, streams, lakes and oceans. Thus, with agriculture based on plowing and other tilling, depending on the climate and method, about 1mm of soil will be eroded each year. The forest, which in many places has been the starting point for agriculture, often has around a meter of soil, so a quick calculation will show that after 1000 years of agriculture, humans will typically leave a desert behind them.[9]

Desert areas can be spread by sand drift, where sand blows in and settles on agricultural land that could otherwise still be cultivated.

At the moment, worldwide we are losing 25 billion tonnes of soil per year, which corresponds to 3.3 tonnes of soil per person per year on average. This is because deforestation is increasing globally, as well as inappropriate cultivation methods that use tilling.[10]

The soil catastrophe and climate change

Climate change is most often presented as a result of human use of fossil fuels. Therefore, it may seem surprising that a significant part of humanity's total historical CO_2 emission has occurred through the loss of carbon from soil and plants when we have cut down and burned forests, cultivated steppes and drained wetlands for farming. It is estimated that deforestation and agriculture over the last 6,000 years have resulted in a loss of carbon into the atmosphere of 349 billion tonnes. Over half of this loss has occurred since 1850. The burning of fossil fuels has primarily occurred since 1750, and this has overall resulted in the release of 483 billion tonne of carbon into the atmosphere.[11] Today, fossil fuels are by far the largest contributor to human-made climate change.

Plowing and loss of carbon

Plowing is a completely unnatural process where huge areas of bare land are created without vegetation to make a seedbed for annual crops. When plowing, we use enormous amounts of energy to work against nature, which will constantly try to repair the bare soil and, if allowed, return it to forest.

When plowing, the top 18–20cm of the soil is turned over year after year; in this way the soil is excessively aerated, which means that an unnaturally large amount of air enters the soil. The oxygen (O) here will combine with the carbon (C) in the decomposed plant material and rise into the atmosphere as CO_2. Plowing therefore reduces the carbon content in the soil.

Plowing also destroys the network of fungal roots (mycorrhizae) that naturally grow in the soil and that are important for the storage of carbon in the soil in the form of humus. The fungal roots are also important because they help the plants absorb nutrients from the soil.

In agricultural land with plowing and cultivation of annual crops, the content of organic matter is 1–2%, while in a permanent grassland that has not been plowed for many years, it is around 5–10%.[12]

By switching to no-till farming systems and the cultivation of perennial plants, we can put the carbon back into the soil. This is an important element in permaculture's strategy to counteract climate change, something we return to in the section 'The Earth must be reforested'.

Salt deserts

In areas where irrigation is intensive, the soil can become salty. When the bare soil is irrigated with groundwater, which contains salts, an unnatural amount of salt will be added to the surface of the soil. This can be further enhanced when there is a lot of evaporation from the bare soil, and water from deeper layers is thereby drawn upwards. Large formerly lush areas in (e.g. Australia) are today barren salt deserts – and since very few plants can grow in salt, it is rather difficult to restore such soil.

Both erosion and salinization are further enhanced by global warming. Over 10% of the world's cultivated land is affected by salinization,[13] while as much as 52% of the world's agricultural land can today be classified as degraded.[14]

Loss of biodiversity

The biological diversity of animals and plants is disappearing at a rapid pace. There are as many species becoming extinct now as when the dinosaurs were wiped out. An animal species that becomes extinct never comes back. The loss of biodiversity is due to natural areas being increasingly replaced by monocultural forestry and monocultural agriculture, as well as infrastructure such as roads and cities. At the same time, nature is put under pressure by climate change and pollution from nutrients, pesticides, chemicals etc.

The various plant and animal species depend on each other in a complex network. If we look at the population of insects, it has declined sharply in recent years. This is alarming, especially because 80% of all plant species depend on pollination by insects and will become extinct in the absence of pollination. The plants form a food base for a lot of animal species, whose populations are also decreasing. Animal species that feed on insects also collapse, and then it is not difficult to imagine what happens to the rest of the food chain...[15]

The energy catastrophe

Climate change

The CO_2 level in the atmosphere rises both because we burn fossil fuels and because CO_2 is released when forests are felled or agricultural land is degraded. The increase in CO_2 and other greenhouse gases in the atmosphere means that more of the heat that radiates from the Earth is reflected back to the Earth's surface, thereby increasing the temperature here.

The pre-industrial level of CO_2 in the atmosphere was 278 ppm (parts of CO_2 molecules per million molecules), but at the time of writing the level has exceeded 427 ppm.[16] The increase in CO_2 in the atmosphere is occurring much faster than with natural climate changes earlier in the Earth's history. A safe level for CO_2 in the atmosphere is believed to be 350 ppm,[17] i.e. lower than the current level. This means that we need to not only reduce our greenhouse gas emissions, but that we must remove more CO_2 from the atmosphere than we emit, in order to keep the situation under control.

If we are to stay close to the target of a maximum temperature increase of 1.5°C, and only exceed this briefly, humanity must reach net zero CO_2 emissions around the year 2050, followed by net negative CO_2 emissions.[18] If, on the contrary, we continue as in the International Panel on Climate Change's (IPCC) scenario for very high emissions, we will reach a 4.4°C temperature increase as early as 2100.[19]

Already today we are experiencing a global temperature rise of 1.5°C,[20] which results in extreme weather in the form of heat waves, hurricanes and torrential rain. There is an increasing number of catastrophic wildfires; glaciers are retreating; sea ice is reducing; sea levels are rising, and coral reefs are dying. These effects will increase in strength as the temperature rises further.[21]

A factor that makes the situation even more dangerous is that climate change creates self-reinforcing effects. An example of a self-reinforcing effect is that retreating ice will expose land and sea that is darker in color than the ice and will therefore absorb the sun's rays rather than reflecting them. This results in higher temperatures, which in turn results in increased meltdown.

Other self-reinforcing mechanisms are that when forests burn, CO_2 is emitted, and when permafrozen areas melt, methane and CO_2 are released. This can put us in a situation where the climate changes that we as humans have initiated can get completely beyond our control.[22]

It is still possible to secure our future by limiting global temperature increases to between 1.5 and 2°C, but this requires rapid and deep reductions in humanity's emissions of greenhouse gases.[23] In order to achieve such reductions, it will require, as was stated in the IPCC report from 2018, 'rapid, far-reaching and unprecedented changes in all aspects of society'.[24]

We cannot wait for technological developments to possibly solve the problem, because by then it will be too late.

The organization catastrophe

The last catastrophe is that neither on a national nor on a global level have we created an organization that promotes the regeneration of natural resources and counteracts climate change and thus works with permaculture ethics. In many cases, this is due to short-term thinking and the defence of the privileges of certain groups and not others. It may also be because the policy is fundamentally based on a different ethic than that of permaculture.

This does not mean that political decisions never promote care for the Earth and people and fair share; because they often do. But this does mean that there is a large gap between what should be done and what is actually being done.

In a situation where we are threatened by climate change, which risks causing our society to collapse and is a threat to the continued existence of humanity, an appropriate reaction would be to initiate an all-encompassing restructuring of society and our economy, such that the problem is solved. Getting it solved is a real possibility if we use the many carbon-storing permaculture techniques. The long-term survival of humanity and all other life on Earth must of course have the highest priority; anything else reveals a lack of overview and a disregard for responsibility.

In a situation where the necessary responsibility is not taken at the societal level, it is left to the individual to behave ethically as a political consumer. But problems are not solved that way as the ethical consumer is consistently punished financially for their choices through higher prices on the things that take care of the Earth and its people.

Basically, fossil fuels are far too cheap

in relation to the damage they cause, and this means, for example, that it is more expensive to travel by train than to fly, even though the climate impact of air travel is on average six times as high as traveling by train. Also, the fact that organic and fair trade foods are more expensive than conventional goods when we buy them is a direct economic incentive to not take care of the Earth and people. Of course we must try to behave ethically and responsibly as individuals, but it is a political responsibility to organize the economy in such a way that it encourages good ethical choices.

On top of the economic misorganization of society comes a number of other societal conditions that either do not support or directly hinder us when we try to create permaculture solutions. These include what the education system supports or does not support, legislative barriers in relation to ecological construction, waste-water management and the subsidy rules in agriculture, which predominantly support industrial practices.

A similarly large gap exists between what should be done and what is actually being done in relation to how we relate to the suffering that daily affects the world's poor, especially in the developing world. With one hand, we give alms in the form of development aid. With the other, we help to maintain unfair trade conditions that prevent developing countries from starting their own businesses, for example, the imposition of tariff walls when agricultural products are sold to the EU and the USA or prohibitions from the World Bank via tariffs on imports when developing countries want to financially protect their core industries, with the result that they cannot develop and achieve competitiveness.

The organizational disaster therefore ranges from the way we live and are organized in local communities and nation-states, to how the world's international organizations are structured. It is beyond the scope of this book to go into depth about national and international organizational problems, so we will therefore simply refer you to the book *Occupy World Street* by Ross Jackson, which provides a detailed analysis and strategy for action.

Vision engineers

The problems described as the five catastrophes must partly be solved at the political and organizational level, and one way in which we can make a difference is by engaging with the established political system. We can also make a difference on the organizational level by criticizing existing conditions, politicians and large companies and by trying to make a positive difference through various forms of political activism that can focus on the problems and thus create political change.

But the problems are also deeply rooted in our culture, and it is difficult to rebel as long as we ourselves are part of this culture. After all, we have the politicians that the population elects.

Therefore, a completely different way in which we can push for change is to use the power of example and start thriving permaculture projects in our local communities. With permaculture in practice, we can show that another world is possible. New people who come to experience it can see it, hear it, smell it, taste it and feel it, and they can talk to people who are in the middle of it – in the middle of the permaculture society of the future.

With our knowledge and our practical examples, we can, by leading this transition, be 'vision engineers'. We can create new positive ideas and stories about the future and share them with our world. This is precisely a central strategy for the transition movement, a movement that, with its origins in the permaculture movement, works to transform entire local communities so that they can address the climate crisis.

Our ideas and visions for the future help to guide the choices we make. Right now, many people cannot imagine how we can create an attractive lifestyle that takes care of the Earth; they are overwhelmed by environmental and climate problems and end up in apathy. They lack concrete images and examples – and we, as vision engineers, must work to ensure that more people see the lush, simple, local and healthy path towards a resource-regenerating and carbon-storing lifestyle.

Political activists and vision engineers are not opposed to each other but can influence the world in the same direction, just on different levels. And there is also an overlap because there is a great need for our politicians to have new visions of the future that include permaculture so that a positive dynamic can develop between the political leaders and the permaculture grassroots.

We need politicians who can see the potential of permaculture in the green transition and who help permaculture projects advance legislatively and financially, instead of focusing exclusively on centrally managed projects that are carried out over the heads of the population.

The more people who embrace a permaculture vision, the more we'll turn that vision into action – building real-world examples and voting in ways that bring it to life.

The primeval forest as a model

In order for us in permaculture to use nature's own ecosystems as a model for creating places that meet human needs for food, housing, medicine, heat, energy, etc., we must have knowledge of the original nature.

In large parts of the world, the natural ecosystem is a grassland with migratory animals, but this book is based where there is a temperate climate and so much precipitation that the natural ecosystem is a temperate primeval forest. This is what we will describe. If we as humans had not created cities, agricultural land and production forests, the land in these areas would be covered with primeval forest. So, what does such a primeval forest look like?

Several layers

The primeval forest consists of several layers. On the forest floor there are herbs, withered leaves, dead wood and mushrooms – the ground is covered with living and dead plant material. Above this are shrubs and young trees, which are protected from the wind and drying out by the fully grown dominant trees in the upper canopy layer. Vertically, there may be a layer of climbing plants.

This stacking of layers utilizes sunlight for photosynthesis from early spring, when the forest emerges from the ground upwards because the light then reaches the forest floor, all the way to late fall, when the leaves fall from the canopy layer. A deciduous tree reproduces the weight of its own trunk in the form of leaves many times during the tree's lifetime, and the other layers below the tree also photosynthesize. In this way, the forest provides a higher 'yield' than cultivated land, also called net primary production.[25]

Many niches and species

The forest is full of niches that are exploited by different plants and animals. For example, when a tree falls and the root is torn up, a hole is created, which often becomes a moist area. The tree trunk becomes a habitat for beetles and fungi. The root above ground becomes a place for mason bees, and later when the root has rotted away, a pile of stones and soil is left behind. Many animals depend on all these different niches being present within a small area.

A fallen tree will also allow light to reach shrubs and herbs – it will create an edge in the forest between clearing and forest – and the edge between the different elements is often the lushest place because the many elements mean that nothing is missing in plant growth or animal life. Another lush edge can be along the shoreline of a forest lake with aquatic plants, fish that breed, animals that drink and shrubs and trees that use the water and the increased sunlight of the area without trees.

All in all, the forest contains a great diversity of both different species and ecological niches.

Life underground

There is also important life underground. Roots from trees, shrubs and herbs draw up water, nutrients and minerals from deep layers. Fungi found on the forest floor are often mycorrhizal fungi, and they form a symbiosis with the plant, where the fungus, which cannot photosynthesize, receives sugar from the plant and in return supplies the plant with water and nutrients (see more under 'Nutrients in forest gardens and agroforestry' (p.111).

Organic life in the soil, with the earthworm as the largest and thus most visible indicator, brings the withered leaves down into the soil and converts them into humus and nutrients that the plants can absorb. All in all, the forest is a self-reliant system that does not need energy, water or fertilizer supplied by humans. The forest is a resilient system that can withstand periods of drought and heavy rainfall.

Beneficial relationships

Often, nature is portrayed as a place where plants compete with each other and animals compete with other animals, which of course does happen. But as described above, nature is full of many more beneficial relationships between the different elements. It is these beneficial relationships that we as permaculturists must emulate in order to achieve resilient and highly productive systems.

Mosaic structure

A tree rarely falls alone but will often take several trees with it, creating a clearing. The clearing will make room for young trees that will grow up simultaneously. This will create a mosaic structure, where in some places in the forest there is a clearing, in some places there is young forest, in other places there is fully grown forest, and in other places there is old decayed forest.

All of these different stages of forest succession are the model for permaculture. The clearing is a model for the grassland with ruminants; the young forest for the forest garden; the fully grown forest for nut forests with tall trees; the old forest for mushroom cultivation of decomposer fungi grown on dead wood, etc.

Where are the annuals?

But where in nature's ecosystems are the annual plants that are currently grown on most of the world's agricultural land? The annual plants only have a small place, for example where a fallen tree or a pig has briefly created bare soil, an open wound in the otherwise always covered forest floor. Here the annual plants come in as pioneer plants, also called repair plants, and ensure that the soil is quickly covered, after which the perennial plants will spread to this place. The common feature of the annual plants is that their seeds are very rich in quick energy in the form of carbohydrates and thus germinate quickly. Many of the annual plants, such as cabbage and beets, also come from coastal regions where the waves from the sea often create bare soil.

The fact that these annual plants, which are so rare in nature, are now cultivated over an enormous area is the reason for the loss of biodiversity and a long list of health, environmental and, not least, climate problems that global society is struggling with today.

The primeval forest has many layers: shaded and open areas in a mosaic structure. The forest floor is full of dead wood, which creates habitat for fungi and insects. Suserup Forest, Denmark.

Large animals and prehistoric humans

Large animals in the forest will reinforce the mosaic structure because they graze and eat young trees, so a light-open area will remain open longer. But which large animals are they? Today, for example, in northern Europe and North America, we encounter only a few large animals in the forests. However, archaeological finds show that the primeval forests of the past, before the arrival of humans, were richly populated by large animals such as forest elephants, mammoths, woolly rhinoceroses, hippos, aurochs and wild horses.

These animals have presumably been exterminated by humans hunting. By studying the Scandinavian kitchen middens from the Mesolithic Stone Age, we can learn that the first kitchen middens are full of bones from large animals and that the animals they ate became smaller and smaller throughout the Stone Age, ending up with large quantities of oysters.

That humans affect and degrade the nature around them is not a new phenomenon. The primeval forest is therefore a more varied and open forest than the forests we know today.[26]

A tree has fallen, and a wet and light-open area is created at its root. Here, the rare bare soil in nature appears – a small catastrophe. Annual plants are rare in nature, but they quickly come in as repair plants and cover the bare soil before perennial herbs, shrubs and trees take over.

Permaculture – the manipulated primeval forest

The hunter-gatherers' diet and probably also working hours have therefore changed in relation to how rich the resource base has been in relation to the size of the population. In a densely populated world like today, we would not be able to feed ourselves in a primeval forest.

So the primeval forest with all its useful connections is our model, but we must manipulate it greatly so that we achieve highly productive systems where we can be 'hunters and gatherers' and harvest nuts, berries, fruit and vegetables and possibly keep animals from which we can get meat and eggs, etc.

Using the primeval forest as a model, we will gradually regenerate all the degraded natural resources and create labor-saving systems. From nature's ecosystem, we can derive the permaculture principles.

Permaculture principles

The principles of permaculture are the very bedrock of permaculture. All the concrete methods and examples detailed in the practical sections of this book function as examples of how the principles can be applied in practice.

These principles are general principles, and they can be applied to permaculture projects anywhere in the world. Overall, they are about using nature as a model; about working *with* nature instead of *against* it. In this way, they are reflective of the structure and way of functioning of a primeval forest.

The principles have been developed on the basis of studies in ecology, environment, energy conservation and landscape architecture. Their purpose is to help us see the big picture as we work to create designs that care for the Earth by regenerating natural resources and that care for people by saving labor and by being highly productive.

Although they are general principles, they are tangible and easy to understand. The principles need not function as a check-list, but more as a toolbox we keep nearby when we create permaculture designs.

In order to create a good design, it is always a prerequisite that knowledge of the permaculture principles is combined with specific knowledge of, for example, gardening and construction and knowledge of local climate and cultural conditions.

In the following review of permaculture principles, we'll be following the way they were first described by Bill Mollison[27] and later by David Holmgren[28] and Patrick Whitefield.[29]

There is a lot of overlap between the different versions of the principles. However, like Whitefield, we have formulated the principles so that they relate directly to the site and how we design for the site. We have thus omitted the attitudinal principles relating to how we organize ourselves as humans and how we approach a task.

Our goal is to produce a version of the permaculture principles that will help us to focus as much as possible on how we regenerate natural resources and counteract climate change in practice. This is because, in our opinion, this is precisely what gives the concept its importance and relevance.

The principles of permaculture are presented here in such a clear and practical way that they can help sharpen our focus and equip us with the tools needed to address the climate, environmental, and biodiversity crises.

We'll start the review with the principles that most directly relate to the way a primeval forest functions and then move on to the principles where the imitation of the primeval forest is more indirect.

Natural soil

No-till

In permaculture, we strive to create the same soil conditions as in a primeval forest, i.e. we avoid tilling the soil by plowing, digging or rotavating, as much as possible. By avoiding this, we also avoid the high energy consumption and over-oxygenation associated with turning the soil.

Permaculture principles
Beneficial relationships
Key planning tools
Multi-dimensional
Relative location
Natural soil
Catch energy
Diversity
Small scale
Wholes
Ethics
Earth care
People care
Fair share
SUNE WATTS
SUNE WATTS

Covered soil

Just like in the primeval forest, we endeavor to keep the soil covered with either dead plant material or living plants. This provides optimal conditions for the regeneration of a healthy, living topsoil with a high content of organic material – a soil that can provide the basis for abundant plant growth.

Perennial plants

In addition to not tilling the soil, permaculture utilizes perennial plants such as herbs, perennial vegetables, shrubs and trees rather than annuals wherever possible. This further reduces labor and energy consumption because we don't have to sow or replant every year. Perennial plants are also generally better at keeping weeds at bay than annuals because the soil is constantly covered in vegetation.

Annual crops also have their place in permaculture systems, and they can be cultivated in accordance with the other natural soil principles.

Diversity

Species

Industrialized agriculture is far less diverse than a primeval forest. Just 15 crop species provide 90% of the world's food energy intake.[30] By creating polycultures with a diversity of plant species, or keeping a variety of livestock rather than just one species, we can make the best use of available resources.

For example, different plants in a polyculture can make use of sunlight differently, and if they have differing root systems, they can also utilize water and nutrients at different soil depths. With livestock, cows and sheep graze in different ways, and so in combination they can utilize the grass better than if just one species is kept.

Highly diverse systems are also more resilient than monocultures. Diseases and pests can spread unchecked in a monoculture, whereas in a polyculture there will always be some species that are resistant to the disease or pest. In a year of drought or, conversely, particularly heavy rainfall, monocultures can fail completely, whereas in a forest garden with many different species, there will always be something producing food.

Varieties and breeds

The diversity of plant varieties and livestock breeds within each species is also important. We leave ourselves vulnerable when we base our food security on a few modern plant varieties that may not prove resistant to new plant diseases. Old varieties of our crops are also often better adapted to fertilizer- and pesticide-free cultivation, while older breeds of livestock are generally better at foraging and surviving without concentrated fodder.

Ecological niches

On a permaculture farm, we also want to strive for a high level of diversity in the form of different ecological niches. That is, to have both young and mature forests, meadows and lakes as models for the permaculture systems, as well as niches of unmanaged nature which are not used for production.

Design with multiple dimensions

In a grain field, what we are harvesting is essentially a thin layer about 10cm thick, once a year.

The farming industry aims to increase yields in this type of monoculture farming using artificial fertilizers and GMOs. Permaculture's answer to creating high yields to feed the world is instead to design in multiple dimensions – utilizing vertical stacking, succession over time and fertile edges.

Stacking

Stacking involves planting in multiple layers – just as we see in primeval forests. Perennial herbs at the bottom and then shrubs, trees and climbers, which will photosynthesize at different times of the year and can be harvested at different times. A grassland will never be able to produce as much as the total production of a well-designed polyculture of grassland and trees. Although the production of grass alone will decrease slightly, the production of trunks, foliage and possibly fruit and nuts will more than make up for it.

Stacking can also occur where we mimic the primeval forest more indirectly. This could look like growing shade-tolerant salad plants between tall pole beans in the vegetable garden.

Succession

When we use the dimension of time consciously and work with succession in our cultivation systems, it will further increase yields. For example, when planting production trees for a nut forest, an orchard, or a diverse forest garden, a design can be created where we think about what we can grow between the trees for the first five years, the next ten years and the ten years after that until the trees are fully grown. For instance, we could grow annual vegetables, berry bushes, cut grass for fodder or keep animals while the trees are growing.

The edge effect

The fertile edge between different niches is another dimension we can consciously work to enhance and make productive. In nature, the most productive places are the areas between different ecological niches. For example; the shore of a forest lake, where there is soil, water, sun and shelter, or the edge of a clearing in the forest, where there is the moist environment of the forest but still enough light for nitrogen-fixing trees and shrubs, or for the production of berries and nuts.

In our designs, we can also deliberately extend the edges. We can create a pond that is elongated instead of round and create a long, warm, sunny and moist cultivation bed on the north shore. Or, as in a forest garden/food forest, we can plant the trees with ample spacing so that the entire food forest becomes a large forest edge. More indirectly, we can use the edge effect to create a fertile edge between the house and the garden. We can plant heat-loving plants against a warm south-facing wall. The plants will get warmth, shelter and sunlight in this edge between the niches.

Beneficial relationships

In permaculture, designs are made where beneficial relationships are created between the various elements, just as the primeval forest is full of beneficial relationships. We say that all elements require some inputs to function and that the elements contribute some outputs (also known as needs and functions).

By using one element's output as input to another element in our permaculture system, we create beneficial relationships and achieve the best possible use of resources, optimized production and avoidance of waste.

In a large design with many elements, an input/output analysis can be done to clarify what goes in and out of the different elements. We should then aim to minimize our need for external inputs and use the elements' outputs elsewhere in the system instead of sending them off as waste – we are creating closed loops.

By designing a home with a composting toilet and a reedbed system for gray-water treatment, we can make the outputs from the house usable elsewhere. Diluted urine from a composting toilet can be used as fertilizer under fruit trees, and composted feces can be used as fertilizer in a coppice woodland (check regulations in your area). Wood from the coppice can provide an input to the house in the form of firewood, and ash from the wood stove can act as an output that then returns to the coppice woodland as fertilizer. The gray water, which has been cleaned in the reedbed system, can flow into a larger bed in the garden, where more reeds are grown to be harvested and used for mulching. Circuits are made between the elements.

Each element has multiple functions

In order to maximize the number of beneficial relationships, the elements chosen for the design should have as many functions as possible. The idea is to make use of not only the main function of an element, but also to recognize its secondary utilities. By seeing the full potential of plants, animals, buildings, etc., we maximize the yield of the system as a whole.

We might want a stove in our house with the main function of creating warmth, but we can also consider one or perhaps several other functions. We could include hotplates or an oven for cooking, heat our water, make biochar and collect nutrients for use in cultivation by cleaning the smoke produced by means of a wet scrubber, in much the same way as the ash from the ash box can be used as fertilizer.

Chickens are another example of an element with many useful functions. See the list on input and output in the section on chickens, p.73.

Each element is supported by several other elements

At the same time, we can design so that every important function is supported by multiple elements. The most resilient system is achieved if essential needs such as food, water, heat and income are met in two or more different ways.

The heating requirements of a household could be met with the use of a combination of passive solar heating, a wood-burning stove and a heat pump. In the event of a power outage, the heat pump wouldn't work, but you could still keep warm on a gray day by firing up the wood-burning stove.

We can cover our financial requirements mindfully by drawing from several sources, e.g. various forms of agricultural production, tours of our permaculture site, pedagogical work on the site, a part-time job, handcrafted wooden or woollen products and so on. If one source of income should fail, we still have the others.

Relative positioning

In order to form beneficial relationships, things need to be placed sensibly in relation to each other. By moving the different elements around on a design map, we can place them well right from the start of our permaculture project, which is much easier and much more manageable than moving things around in real life.

Let's imagine a design that includes the elements of a home, a greenhouse, a pond, a vegetable garden and a duck yard.

We can choose to build the greenhouse directly adjoined to the house with a door in between instead of building a separate greenhouse in the garden. This shields the house from wind chill and allows us to let warm air from the greenhouse into the house during the heating season, thus saving energy for heating. Conversely, the heat loss from the house at night will help keep the greenhouse frost-free for a larger part of the year. At the same time, it becomes much more convenient to take care of the cultivation in the greenhouse, especially in the case of bad weather.

The pond can be placed south of the greenhouse so that the low winter sun can be reflected from it up into the greenhouse, providing extra warm air to heat both the greenhouse and the house, as well as allowing for plant growth later in the fall.

The kitchen garden might be placed lower in the landscape than the pond, allowing us to water with the force of gravity and thus without the use of a pump.

We place the duck yard near both the vegetable garden and the pond so that the ducks have their necessary access to water and can swim in the pond. They can help keep slug levels down around the vegetable garden and are allowed into the garden to eat slugs during the winter months and also at times when the vegetables are large and robust enough not to be damaged.

By placing things in the right place in relation to each other, beneficial relationships become practical. For example, if the duck yard isn't next to the vegetable garden, it won't be practical to let the ducks into the vegetable garden.

Small-scale intensive systems

This principle should in fact be called appropriate scale, but in our centralized, large-scale society, the current normal size is often far too large.

The principle of small-scale applies to a wide range of structures in our society. By bringing functions back to local communities, we reduce energy consumption for transporting people, food, water, wastewater, organic waste, etc. Here we will use agriculture as the main example of the principle.

High yield

The principle of small scale is a prerequisite for many of the other principles to become practical. And it's when we design with these that we achieve a highly productive and labor-saving system that uses nature as a model.

There is a widespread belief that large-scale farming is the most efficient way to produce food. However, this is only true if you define efficiency in terms of a small number of farmers working on industrial farms being able to produce food for many people.

According to a metric of the amount of food produced per unit of land, or the amount of food that can be produced per unit of energy consumed in agriculture, small-scale agriculture and gardening are the most efficient.[31, 32] This is because on a small farm or in a garden, we can give each piece of land more attention and care. The energy efficiency of small farms is also related to the fact that at least 75% of the world's small farms are run organically without the input of artificial fertilizers and pesticides.[33]

Small-scale and diversity

Diversity, in particular, often disappears completely when we go large scale. On a big farm, we need to keep things simple in order to maintain an overview, creating large monocultures, or possibly simple polycultures with two or three species, where production is suited to the use of large machinery.

On a small farm or in a garden, several of the permaculture principles can be put into practice. For example, we can create sophisticated polycultures with covered soil and a high level of diversity; we can fashion beneficial relationships between animals and crops; we can make use of stacking and fertile edges; and we can utilize different parts of the landscape to their best advantage, rather than creating large uniform fields of monocultures.

Reflecting on small-scale systems

The idea of small-scale is also where permaculture breaks away the most from societal structure and the common assumption that the bigger something is, the more efficient and professional it is.

Structural developments in recent decades have resulted in more and more farmland globally being transferred from small-scale farms to industrialized agriculture, and as large-scale agriculture produces less food, this is a threat to global food security.[34]

Permaculture proposes that both intensive gardening in home gardens and permaculture farming oriented towards the sale of products are important for the green transition of our food system.

On a farm that produces food for purchase, growing systems are scaled up in comparison to a home garden. In a market garden, the no-dig garden beds will often be long and straight, and there may be more employees so that economies of scale can be realized.

A forest garden can also be simplified and expanded into a food forest with a simplified ground cover layer, and then to simplify even further, the same growing principles can be applied to create an alley cropping system, where trees and other crops are grown in rows.

When upsizing, some principles will be dropped; one might neglect to use biological resources for construction, or end up cultivating the soil between trees.

When a system strays a long way from using nature as a model, it is debatable whether it can still be called permaculture. In an agroforestry system consisting of large fields with long distances between rows of trees, where plowing is still carried out, no soil cover used and no local sales made, many of the permaculture principles are not in use, but it is still a positive that trees are incorporated.

Harvesting energy

Energy flows – harvest, storage and utilization

Energy flows through the places we live – for example, in the form of solar energy, some of which is stored in biomass through photosynthesis. There is also energy in water traveling from high places in the landscape to lower ones, and there is energy in wind blowing across the site.

If we do nothing, the energy will pass through our space without us utilizing it. When energy is transformed from one form to another along its path, some of it will always be turned into energy of lower quality, usually heat. This is called entropy.

Our job as permaculture designers is to 'harvest' the energy from these energy flows, possibly creating stores for the energy where it is in the form that is most useful to us, and then utilize the energy in the best possible way.

For example, when we store water high in the landscape in our design, we can use the potential energy by watering relying on gravity alone, and if the height differences are large enough, we can install a turbine and use the energy for other purposes. If we did nothing, the water would run down to the lowest parts of the landscape on its own, and the energy would not be utilized.

We can plant a Siberian pea tree that produces seeds with a high protein content. In this way photosynthesis has converted a certain amount of solar energy into chemical energy, which is stored in the seeds. You could say that we have 'harvested' the energy. If we then feed the seeds to a chicken, a small part of the energy in the feed will be used to produce eggs and meat and utilized to keep the chicken moving. The majority will become body heat.

Our job is to make the best possible use of these forms of energy rather than letting them go to waste. For example, the chicken's kinetic energy can be utilized by letting the chicken roam where we want to construct beds or sow a larger field crop, scratching the soil to clear it of weeds.

We can capture the chicken's heat energy if our chicken coop is built in tandem with a greenhouse, which, due to the heat from the chickens, can be kept frost-free for more months of the year.

Other areas where we can strive to make best possible use of the harvested energy is by insulating our homes as well as we can and by driving energy-efficient vehicles. We ourselves or someone else somewhere has harvested the energy and made it available to us, and we need to make the most of it.

Embodied energy

Embodied energy is a term that describes the energy used directly to make a material. It can be a man-made material or one that nature has produced. A lot of energy is used in the production of building materials, for example.

Nutrients, such as those found in animal dung or in urine and feces from a composting toilet, also represent an amount of embodied energy and must be utilized in the best possible way.

Like the primeval forest, which retains all its nutrients and circulates them locally, we need to create cultivation systems that retain nutrients and do not lose them through, for example, leaching. This would allow us to be independent of fertilizer inputs, which account for a large part of the energy consumption in conventional agricultural production.

In permaculture, we strive to use materials with low embodied energy and to make things last as long as possible so that we make the best use of this energy.

Biological resources

Biological resources such as plants and animals form an important part of nature's ecosystems. Plants get their energy directly from the sun, and animals get their energy by eating plants. To use biological resources instead of fossil fuels is therefore to use renewable energy.

One example is using beams made from wood that has come into being via solar energy, instead of steel beams. Steel is a limited raw material, and large amounts of fossil fuels are used in its production. Another example is to use nitrogen-fixing plants instead of artificial fertilizers.

Using living biological resources rather than dead ones can be beneficial because they can renew themselves. An example of this is planting a hedge instead of building a wooden fence. The hedge has a far longer lifespan and can also contribute to biodiversity, as well as storing carbon.

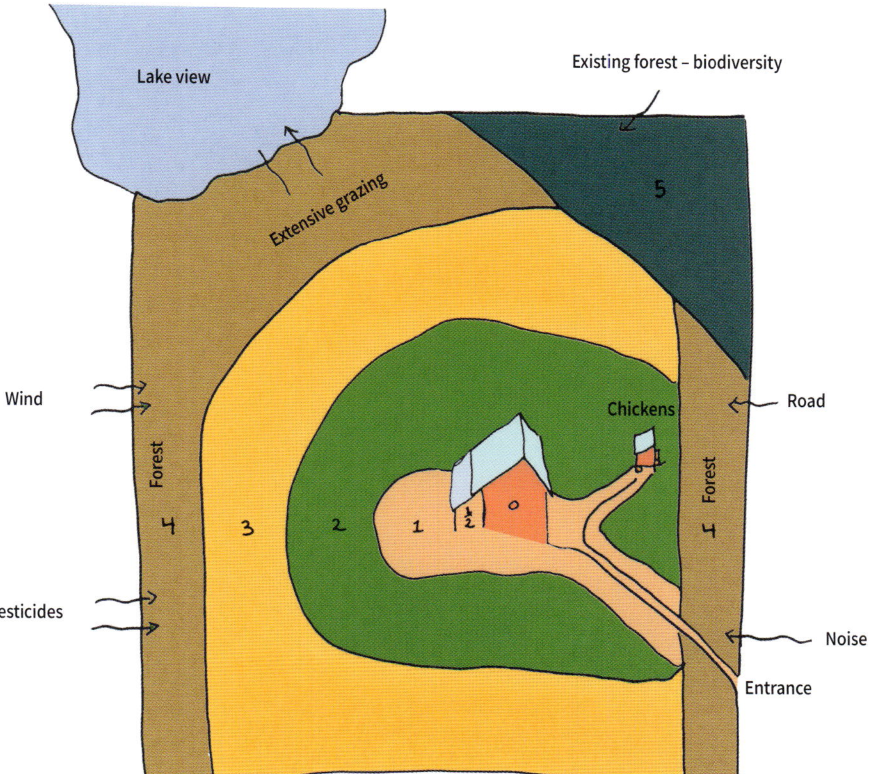

Example of how the connecting paths to the chicken coop and the road, as well as sector influences, such as noise, pesticides, forest biodiversity and lake views, can influence the zone design.

Zone 4 forest stops wind, pesticides and noise. Zone 5 is placed next to the existing forest, achieving an integrated area of wilderness.

What happens in each zone?

0. The house or other center of activity – the place where we spend the most time.

1. Very intensively cultivated area with forest garden, annual vegetable beds, herbs, espaliered fruit trees and fragile plants. Productivity is high, and organic material and fertilizer is usually imported to zone 1 from the zones further away. We come here several times a day and put in a lot of work.

2. Relatively intensively cultivated area with food forest and no-dig beds with main crops; chicken coop, stable and workshops. We come here daily.

3. Farmland, arable farming, grassland/meadow, alley cropping, hardy fruit and nut trees not requiring pruning, small intensive woodlands. We come here periodically.

4. Extensive grazing and forestry, native plant species, export of organic materials. We rarely come here.

5. Wild nature. Here, wild animals and plants are prioritized over production. We only come here to experience nature or facilitate optimum conditions for it. Every design should have a zone 5 because it promotes nature and biodiversity.

Planning tools

The next four principles are not derived from nature as such, but are useful tools. They can be collectively referred to as planning tools because they are used in the design process for the overarching plan.

Zones

The zoning principle is based around the idea that the element that needs the most attention is placed closest to the center of human activity. This saves labor and energy, and intensively cultivated areas are better cared for because they receive more attention. This may sound like common sense, but if we look around our local landscape, it's rarely the practice.

Networks and connectivity

There is often more than one center of activity in a design, such as a house, chicken coop, and workshop. In between, there will be a well-frequented network of roads and paths. A driveway to a public road can also be a center of frequent use. Zones 1 and 2 are often placed along roads and paths that are frequently traveled.

Sector influences

These are influences that come from outside the area you are designing for. The influences either come from a specific direction or affect a specific area (a sector). Examples include wind (dominant wind direction), sun, water that runs across the landscape, noise, pollution, areas at risk of flooding, areas where cold air collects (frost pockets), views (particularly beautiful or ugly) and areas at risk of fire. Soil types are also an influence, and although it's not external, it's important for overall planning. A nonphysical but often very significant influence is government regulation of the landscape.

All these influences can be plotted on a map. By being aware of them, we can

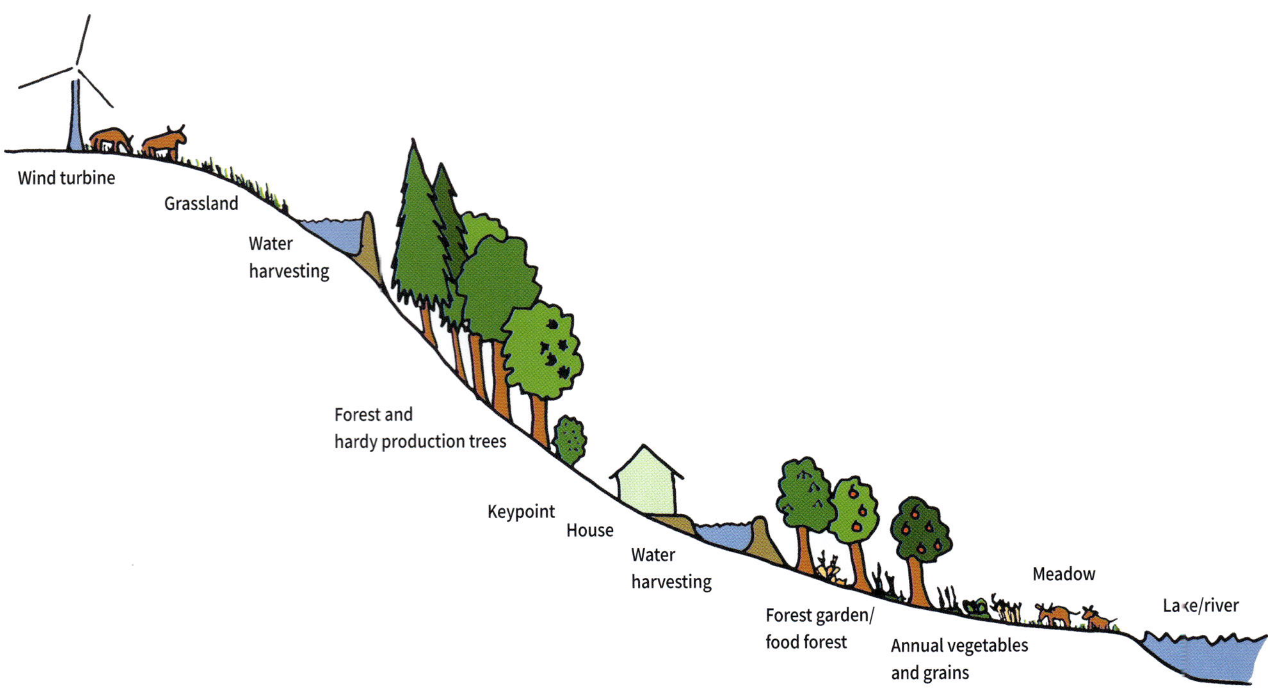

position design elements to best manage these external influences, such as facing towards the sun and shielding from the wind.

Landscape profile

Terrain planning is about placing things in relation to the shape of the landscape. This is especially important on steep terrain.

On an elongated hillside or mountainside, all of the following parts of the landscape profile may be included in a single permaculture project.

The hilltop is exposed to wind. A wind turbine can be installed here, and grassland or forest can be created because they can withstand wind. Annual crop cultivation is an option because there is usually no slope to the land. Water harvesting from this area can provide irrigation to all the other areas.

The center of the hillside is usually steepest. If it's very steep, preventing erosion should be a top priority. A forest could be planted, for example.

Just below the point where the hill goes from concave to convex, you'll find the so-called 'keypoint'. This is a place in the landscape that has many favorable characteristics. It's often sheltered, cold air can flow down to lower areas, and it's high enough to avoid flooding. A keypoint is a great location for a house or a forest garden. There is potential here to collect large amounts of water from the higher areas, and there are still lower areas that can be watered by way of gravity. Keypoints are also very often the point in the landscape at which springs are found.

Just below a keypoint, the slope is gentle and the topsoil is often deep. Annual plants can be grown here without the risk of erosion.

Below the hill, the area is flat, often with poor drainage and risk of flooding. This is a good place for a meadow for haymaking and grazing animals, or for a coppice woodland with willow or alder for fuel production.

Wholes

In an ecosystem, the different parts interact to form a complex whole. In the same way, the whole is both where we start and where we end when we do permaculture design. We look at the wholeness of all the elements of the design and the permaculture principles used. It is the totality of the garden, the house, the people living in the house and their activities that make up the permaculture design. And still, this is only a part of the greater whole of the bioregion, the community and the Earth. Permaculture is thus not a cultivation technique we can practice in a corner of the garden. Everything we do interacts with the Earth, and this is constantly in our consciousness as we work on permaculture design.

Permaculture design

Permaculture is characterized by the fact that a thorough design is always made before the system is established. When we change the positioning of the elements in our environment, in relation to water, soil, buildings and plants to create long-lasting and resource-regenerating systems, we are making decisions that extend many years into the future. It is therefore important that we think carefully.

A permaculture design can help us to create well-functioning and efficient systems and therefore increase the project's chance of success. We can save a lot of labor by getting the elements in order right from the start, rather than rushing out to plant fruit trees, for example, only to find later that they are planted too close together, placed in the wrong location, or that the species can't thrive at the site at all.

We save a lot of labor in the long run by creating a good permaculture design, and that is also how we ensure we are creating highly productive systems.

If we don't create a design, we can, for example, follow some permaculture principles and techniques in part of our garden, such as working with covered soil, perennials, polycultures or stacking, but this isn't really coherent permaculture.

It is in working with the permaculture design that we have the opportunity to change the relative positioning of the different elements, such as the chicken coop, the greenhouse and the vegetable garden, so that we can realize the beneficial relationships that save labor and provide a high yield.

It is also in working with the permaculture design that we use the four planning tools to ensure that we are taking into account the landscape profile and different sector influences, such as frost pockets and wind, in order to work with the landscape and create a solid zone and network design.

The design process

A good design process is characterized by first taking time to observe the area and listen to the people involved in the design.

Only later, when we know the conditions well, do we start making design decisions. A long design process also gives us time to come up with good ideas.

The following describes a permaculture design process in seven steps.[35] Steps 1–3 are the observation phase, where we gather knowledge and observe without making judgments or decisions. Steps 4–7 are the active phase, where we design and make decisions based on our knowledge; next we establish the design and then we evaluate it.

Observation phase

1) Base map

Make a true-to-scale map of the area. The map should only show the things that already exist on the site. Include buildings, access roads, divisions (hedges, walls, fences), plants, water (ponds, streams, taps, downpipes, etc.).

Draw the landscape profile (contours) on the base map or use an overlay map. You can also do a cross-sectional drawing of the area in profile.

Hørhaven in Denmark is 5.5 hectares. The design map shows how the elongated shape of the area has resulted in a clear zone design. Chalotte Vad grows perennial and annual vegetables, fruit, berries and nuts. Chickens, ducks, geese, sheep and pigs are integrated into the systems.

CHALOTTE VAD

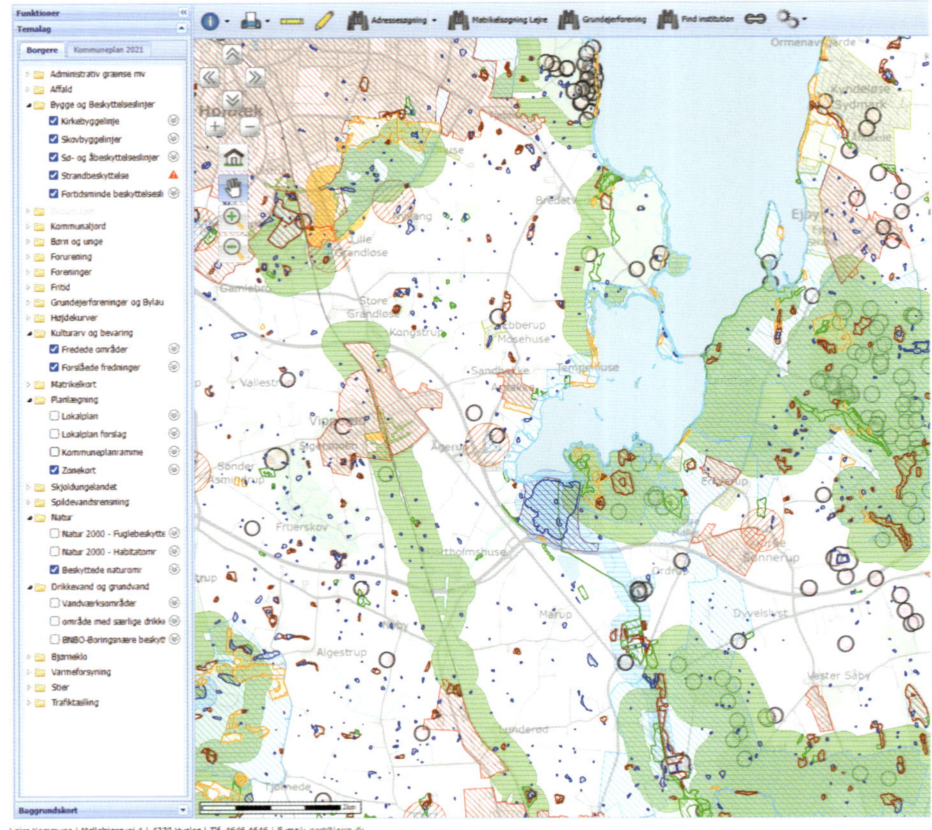

Public regulations affect our permaculture designs. In some countries, official digital maps can be found on the internet. By choosing from the menu, we can see contaminated land, cadastral boundaries, nature and cultural heritage protection, etc. This map from Denmark shows large red circles indicating church protection lines; the small circles are protected ancient monuments. In addition, there is river and coastal protection. This protection prohibits the planting of trees. In light of the climate crisis, preventing permaculturists from planting trees on degraded plowland is outdated legislation. If we're purchasing new land for a permaculture project, it's a good idea to research public regulations before we buy.

2) Get to know the site

Gather information from both available data and your own observations. Some information can be written down, and some can be drawn on a copy of the base map.

Find old maps or photos of the area. Maybe the neighbour also knows something about the history of the area.

Take a walk along the edge of the area. Take a walk at night and in heavy rain. If possible, get to know the site in all four seasons.

Relevant information can include the regional climate conditions, the microclimate, the wet and dry areas, the soil conditions and the plants and animals on site, including pests.

What are the needs for water, and where can we get water? What cables and pipes are in the ground?

Record what buildings there are and their condition. Do the same for roads, paths, bridges and gates, as well as information about the traffic conditions and public transport.

The spade test is one of the tests we can perform in the observation phase. It tells us about the condition of the soil. For example, are there deep roots or is there a hard, compacted layer of soil just below the plowed soil?

We are using two spades to extract an undisturbed clod of soil. First one spade is inserted into the ground. Then a hole is dug in front of the first spade. The clod of soil is cut free at the sides and lifted up.

A good soil will have a fine crumb structure where the soil breaks apart like breadcrumbs. This structure is created by roots, mycorrhiza and earthworm excrement.

Make a sector analysis of external influences that come to the area from outside, e.g. wind and sun, shady trees, noise, poison from sprayed fields, views, conservation areas and other public regulations of the landscape, etc. Draw the sectors on the base map or on an overlay map.

The zone-and-network analysis can also begin with the immovable elements, such as existing buildings and access roads to the area. These zones and networks can change as more elements are placed.

3) Listen to the people

Interview those involved (including yourself, if applicable). What is the overall vision for the site? For a permaculture project to be viable, the vision must be in line with permaculture ethics. The vision provides the overall direction for the project.

What is the state of ownership of the site? Is there a lease, and if so, what are the terms?

What do the people involved want from the site? For example, what do they want to produce? Is it for their own household or to sell? How much income is required from the site?

What recreational opportunities are desired, such as play and socializing? What aesthetic preferences do you have? Which wishes for the site should be prioritized? What are the resources of those involved in terms of time, money, equipment, knowledge, skills and health?

What elements and conditions do you want to keep as they are, and is there anything you want to change or get rid of?

What works well? What works badly?

Are there any elements or relationships that could work better if an element were placed in a different location?

Active design phase

4) Assessment

Review the interviews conducted and assess whether there is coherence between the wishes of those involved and their resources. Then assess which elements could be incorporated into the design. Is there anything that should be deliberately established later or left out altogether?

Look at the maps and information from your site investigation and assess which of the selected elements fits where in terms of zones, sectors, landscape profile and networks.

If it's a complicated design with many elements, it can be a good idea to do an input–output analysis to help us create many beneficial relationships.

List all the existing elements and the proposed new ones. For each element, note down what inputs they need and what outputs they provide. Try to include only the essentials.

Inputs can be space requirements, preferred soil type, microclimate, feed, materials, energy, heat, labor, skills, capital, information, etc.

Outputs can be any physical product, the useful behavior of an animal, an effect on the microclimate, soil improvement, the creation of a habitat for beneficial wildlife, beauty, enjoyment, etc. Undesirable outputs can be pollution, shade, etc.

Look to see if you can create some beneficial relationships between inputs and outputs at different locations in the design. Can anything be moved around to facilitate these connections?

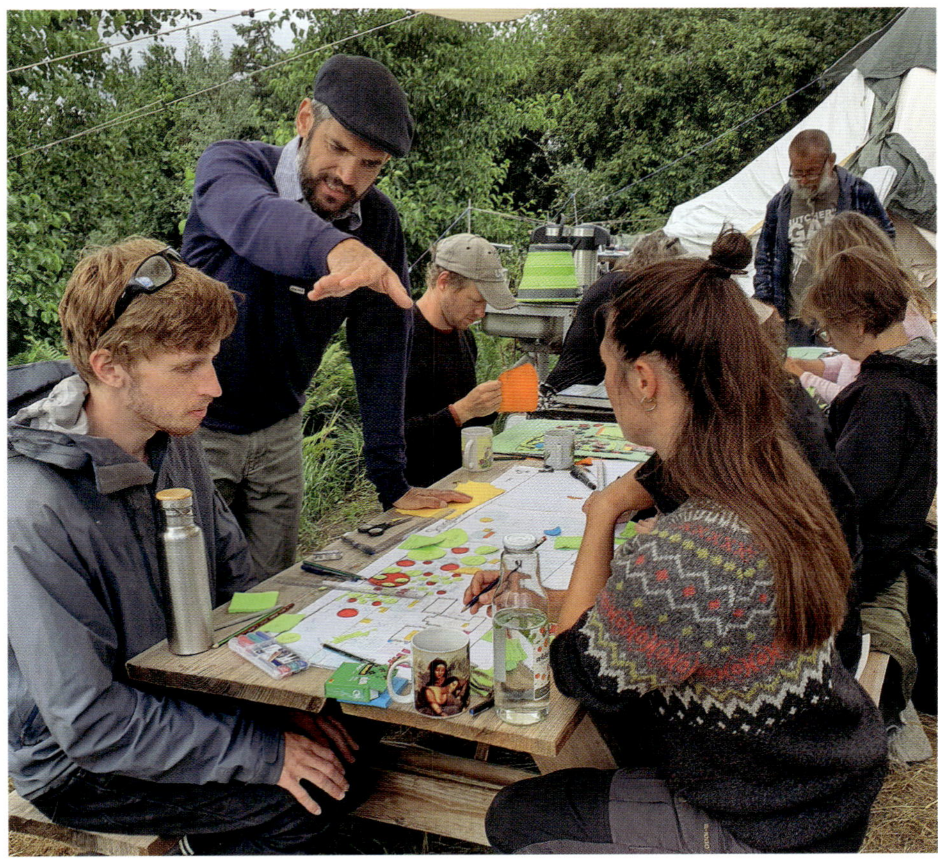

The final phase of the design process involves the placement of trees and shrubs at Skovsgaard on 1.5 hectares. Three generations of children, parents and grandparents live here together. Denmark.

Can new elements be introduced to make use of untapped outputs?

Or are there new elements that could provide inputs that would otherwise have to be imported from outside?

Where can elements be included in the design with multiple dimensions, such as stacking, edge effect and succession?

5) Design proposal

Finalize what the goals/vision for the design are based on your interviews.

Then task yourself with drawing several overview sketches of how the space could be utilized, including only the main elements.

Show on the sketch maps how the zones and networks will look in the various proposals. What beneficial relationships can be formed between the elements?

Choose a sketch to work on and create the final zone-and-network map. From the selected sketch, create the final design proposal, working on more details, e.g., designing polycultures, deciding how the soil is covered and adding diversity.

For permaculture projects with larger production for self-sufficiency and sale, an implementation and maintenance plan, yield assessments and cost and revenue estimates should also be made.

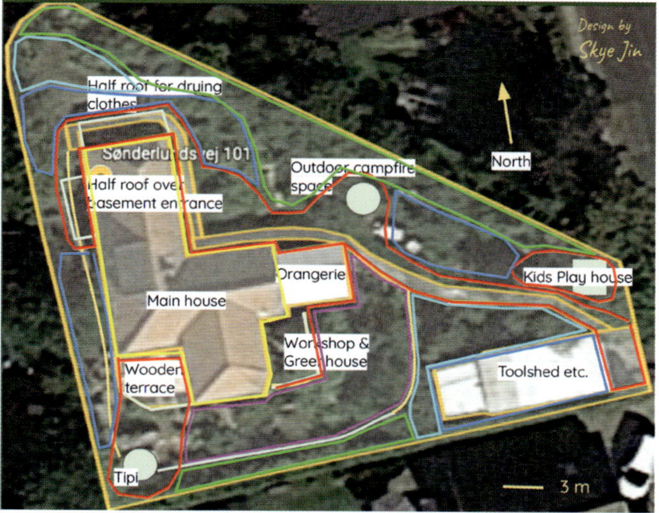

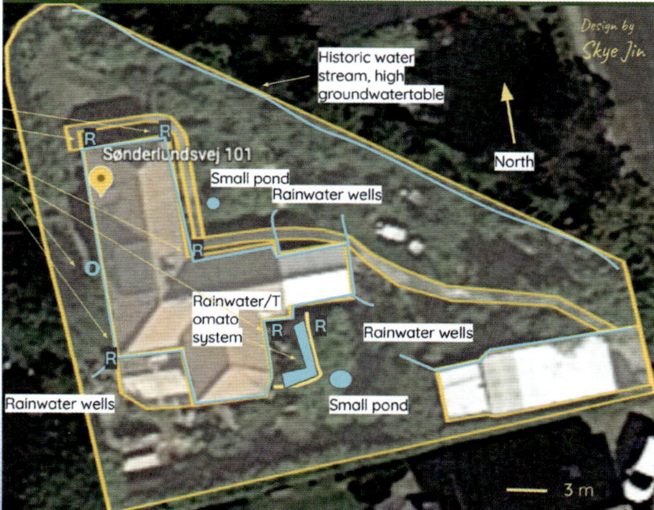

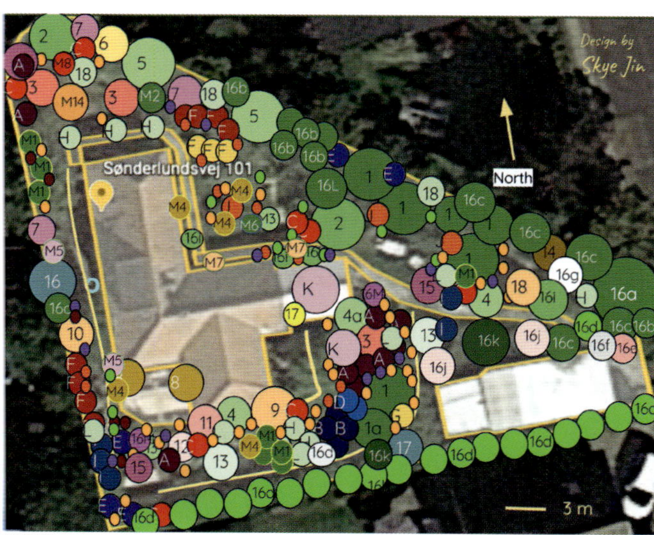

SKYE JIN

Skye Jin has created design maps for her family's house and garden, producing vegetables, fruit, berries and nuts over a long season. She works in different map layers.

Zone design

Water design

Tree and shrub design

In addition, there are also base maps and maps with ground cover layers. Denmark.

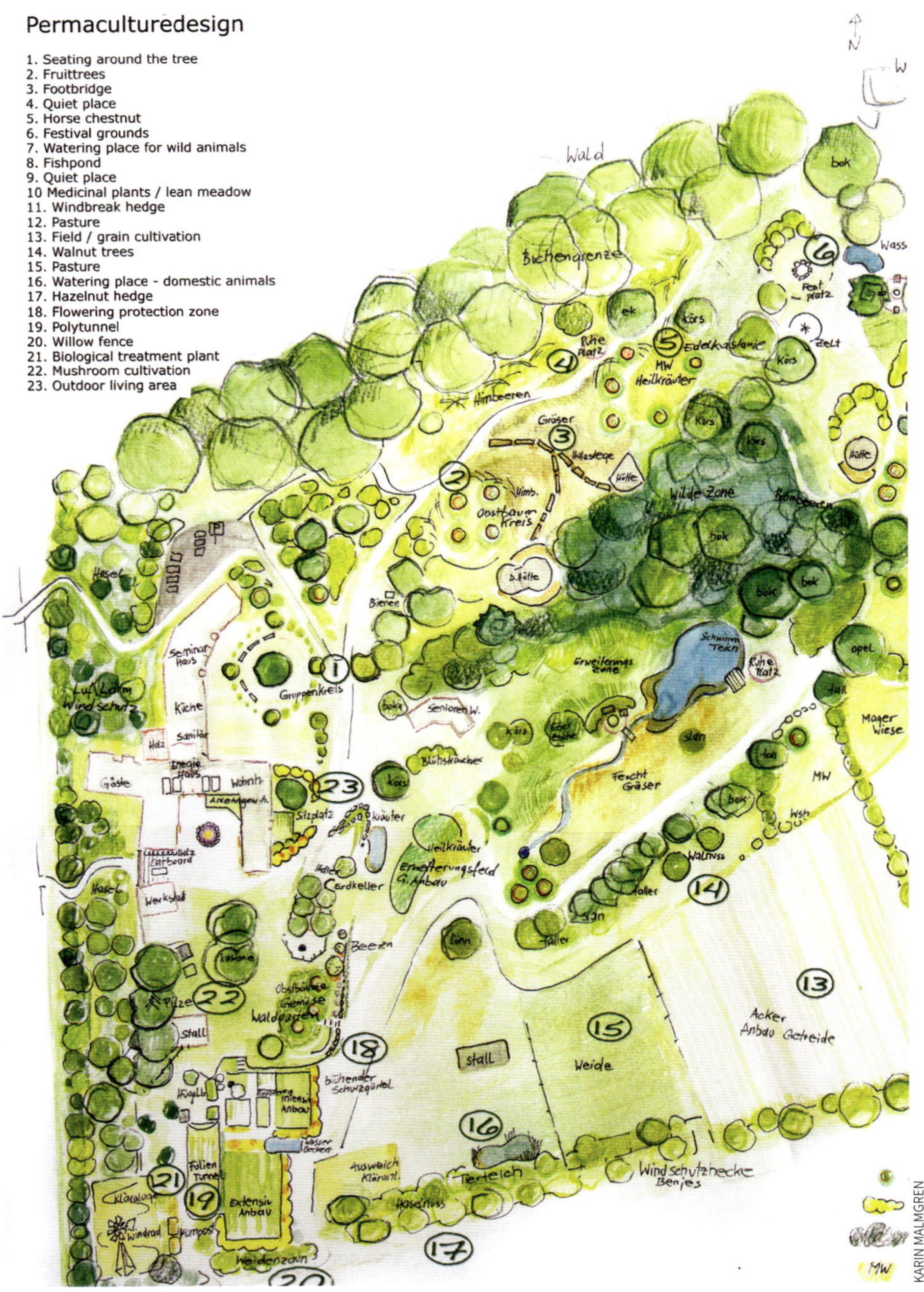

Design map for Ecotopia. Site for 7 hectares, containing animal husbandry, course site, several types of cultivation etc. Sweden.

Presentation at the PDC course of the finished design for Landbruget Underskoven, which is one of six mini farms of 1.5 hectares each in Skovvirke. Today, Lars and Signe Gerlach live in their off-grid straw-bale house and have established their permaculture farm. Denmark.

6) Establishment phase

Building on the design, work can now begin in a focused and efficient manner to bring the project to life.

7) Evaluation

As we gain experience of how the design works on site, there will be elements and aspects we want to change. As an outside designer, it's a good idea to return to the site and see which aspects of the design worked and which didn't. It's time to ask the question, does the project in practice actually care for people and for the Earth? To assess whether a project actually cares for the Earth, it's a good idea to make some specific measurements and calculations, such as measuring the carbon content of the soil and calculating the total ecological footprint.

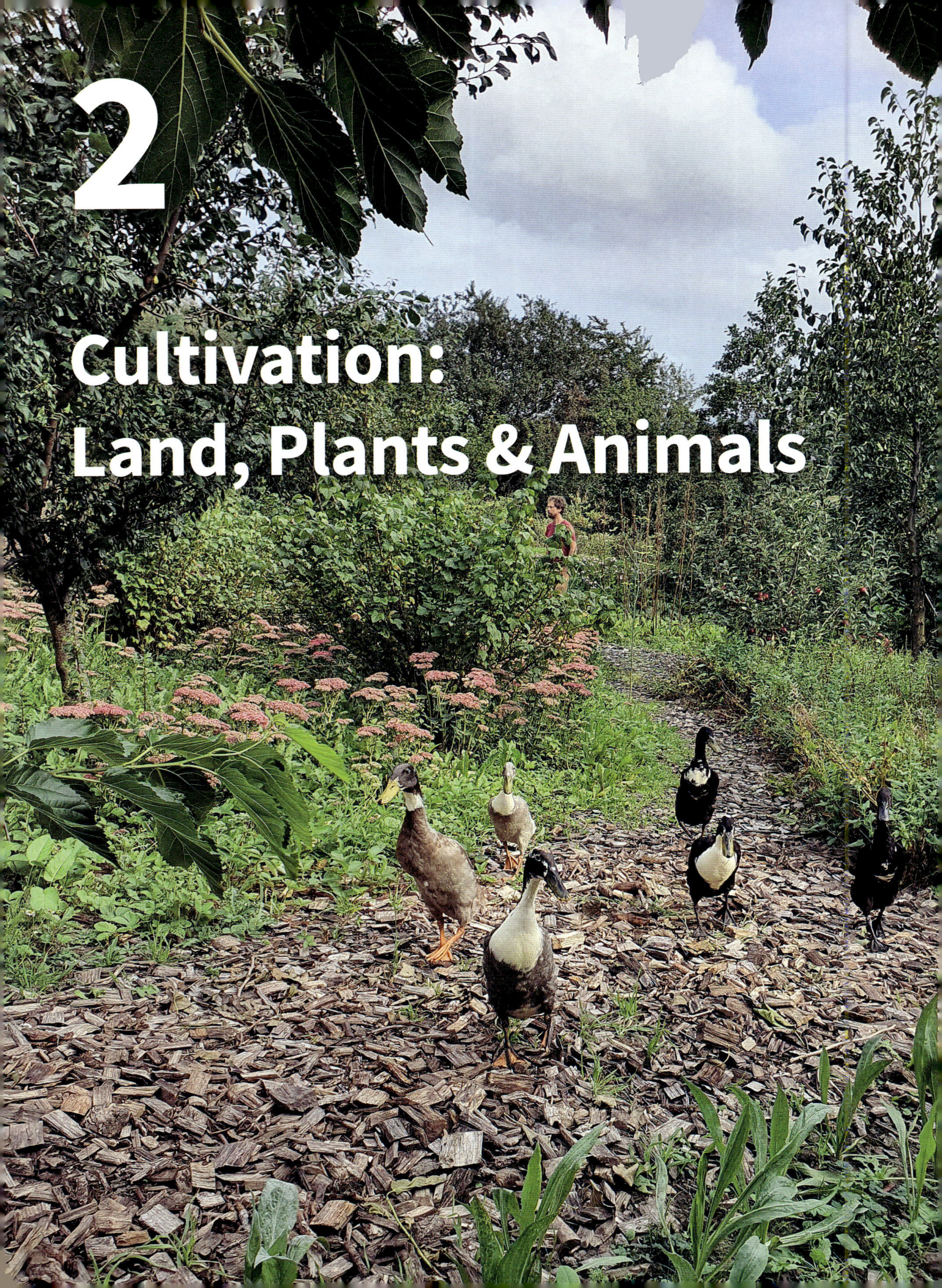

2

Cultivation: Land, Plants & Animals

Visions of Earth's reforestation

Before human intervention, approximately 57% of the habitable land on Earth was blanketed by forest. Today that figure is down to just 38%.[1] A central vision of permaculture is therefore that the Earth must be reforested. By plowing to create bare earth for sowing, we are actively working against nature's process of greening. Transforming fields and deserts into forest will bring about the regeneration of ecosystems and free us from working year after year against nature's process of becoming forest.

Forests that meet the needs of humans

In permaculture we need to work *with* nature and accelerate its continual process of reforestation. This involves a collaboration between humanity and nature to create forests that are intentional rather than coincidental, in doing so fulfilling our needs for building material, fuel, medicine, clothing and, first and foremost, food.

We take the ecosystem at hand as a starting point and tinker with it, planting useful trees, shrubs, climbers, herbs and perennial vegetables.

This can be done in intensive forest gardens with a high level of diversity – comprising perhaps up to many hundreds of different species and varieties – or it can be done in more simplified agroforestry systems where one or a few different productive trees and shrubs are grown in a polyculture with a single ground cover plant, perhaps also with animals grazing under the trees and feeding off the trees' leaves, seeds or fruit.

The food forest is harvested from as early as March, and by September it is laden with apples, pears, hazelnuts, silverberries, perennial kale, English spinach and more. There are large beds with simple polycultures of perennial vegetables – in this way production is scaled up for local subscription sales. Reforest Farm, Denmark.

A whole new landscape

The permaculture vision of food production from forests will bring about an entirely new landscape. Imagine a mosaic structure of different types of forests: groves of tall nut trees with animals grazing underneath, food forests populated with various fruit trees, meadows dotted with pollarded trees providing leaf fodder and other copses producing poultry feed, with egg-laying hens running around underneath. In between these forests are open areas for growing grains and annual vegetables.

Reforestation and carbon sequestration

In the process of reforestation, carbon will be stored in the wood of the trees, in the dead leaves on the forest floor, in dead plant matter in the soil and by the sugars from the trees' photosynthesis. These sugars are sent down into the roots and from there into the mycorrhiza network. The mycorrhizal fungi store the carbon in the soil as humus. Only about half of the carbon stock in a temperate forest is made up of the above-ground parts of the trees; the rest of the carbon stock is found in the soil.[2]

If we convert plowed land into forest, 4–14 tonnes of CO_2/ha/year will be absorbed in the first 80 years of growth, depending on the type of forest.[3] Broadleaved trees and natural recolonization are at the low end, and a thinned Sitka spruce plantation is at the high end.

An interesting question that has been posed is whether we can achieve the same level of carbon capture in a cultivated ecosystem which mimics nature, as in a natural forest. Researchers investigated carbon sequestration in Martin Crawford's forest garden at the Agroforestry Research Trust in southern England, and it was estimated the 25-year-old forest garden stored carbon equivalent to 17 tonnes CO_2/ha/year.[4]

This shows that the cultivation of forest gardens is among the most effective cultivation techniques when it comes to sequestering carbon in a temperate climate, and it in fact outpaces the carbon-storing capacities of forests in their growing phases. The pace of carbon storage in forest gardens is important, as tackling climate change requires a massive effort over a short period of time.

The carbon-storing abilities of a forest decline when it is around a hundred years old. The total carbon storage is likely to be greater in a forest than in a forest garden over their lifecycles as the trees in a forest grow taller.

Rattan Lal is an internationally recognized researcher on the relationship between climate change and Earth's ecosystems. He has calculated that a worldwide transition away from farming practices currently in use on croplands and grazing lands would result in a carbon sequestration of up to 445 gigatonnes of CO_2.[5] This would lower the concentration of CO_2 in the atmosphere

Quicker carbon sequestration in forest gardens than in forests

The rapid carbon storage of up to 17 tonnes CO_2/ha/year in forest gardens is due to a variety of factors:

- Forest gardens comprise a large diversity of plants with varying root systems, which collectively maximize the root mass in the soil.
- Forest gardens contain nitrogen-fixing plants, leading to greater overall growth.
- Forest gardens also contain dynamic accumulator plants, which grow vigorously and deliver carbon deep into the soil.
- Forest gardens are usually more carefully tended by humans than a forest, giving the plants extra favorable growing conditions.

Photo from September. Harvest is complete, and the soil is plowed and sown. Energy from the sun will only be used minimally for photosynthesis until spring next year. There are no niches for biodiversity. The soil is releasing carbon and contributing to climate change.

September in the Agroforestry Research Trust forest garden in England. Pictured here is the modified forest, with many layers and niches for biodiversity. Production is high, and leafy greens, berries, fruit, nuts and mushrooms can still be harvested. Photosynthesis is underway, and carbon is being stored in the wood and underground, mitigating climate change. In early spring, the perennial plants in the ground cover layer will begin photosynthesizing again.

Silvopasture in August. Blekinge, southern Sweden. The large animals open up the forest. Perhaps this is what the ancient primeval forests with many large animals looked like in some places. The grass undergoes photosynthesis as soon as the temperature is above 5°C. The animals also feed on the foliage of trees and shrubs.

by 57 ppm. This represents a vital piece of the puzzle in reducing the amount of CO_2 in the atmosphere down to a safe level where we don't risk uncontrollable, self-perpetuating climate change. It is crucial that this is also combined with vast reductions in greenhouse gas emissions.

All in all, reforestation holds huge potential for how we can counteract climate change. Food-producing forests may be considered relatively safe carbon stores as we will have a vested interest in maintaining a continual production from them, especially in a future with fewer resources available. At the same time, deciduous forests in regions with a humid temperate climate are rarely exposed to forest fires, so in these parts of the world the risk of large quantities of stored carbon returning to the atmosphere is minimal.

Grassland as a model

In arid regions of the world, grassland is the prevailing biome as there is not enough rainfall to sustain a forest. In these places, a grassy plain is the model we should be working with in permaculture.

Although the holistic grazing method described in the section on permaculture farming (p.68) was developed in such areas, it can still be utilized in places where forest is the primary biome and, like reforestation, will store carbon in the soil.

In spring 2016, robust trees and windbreaks were the first to be planted at Underwood and Reforest Farm in Skovvirke, Denmark. Later on, exotics such as peach trees, fig trees and other species that require a warm and sheltered microclimate were planted. Many perennial vegetables require partial shade and were planted even later.

Microclimate

In permaculture we work intentionally with the microclimate. In doing so we can create comfortable outdoor living spaces, extend the growing season and create cultivation systems which are more resistant to drought, strong winds and rain. By improving the microclimate, we can increase the yield of our crops, create greater well-being for humans and animals and make us more resilient to the challenges climate change will bring in the future.

Working with microclimates is, in part, about taking the existing landscape and microclimate as a starting point, which has otherwise been ignored for the last hundred years, where the same crops have been planted all over the world up and down south- and north-facing slopes, wet valleys and dry hilltops.

In many cases the same model houses have been placed everywhere, without considering the direction of the sun or where the naturally warm areas are in the landscape. This is a waste of a lot of good opportunities and results in greater overall resource consumption.

We also design based on the existing regional climate, which we have no influence on, by choosing plants that grow well in the given conditions.

When we do permaculture design, we can also change the local conditions to improve the microclimate. In temperate climates, working with the microclimate will be about creating light, shelter and warmth; while in the tropics it will be about creating coolness, shade, and reducing evaporation.

This section will primarily deal with microclimates in relation to cultivation and outdoor conditions, however microclimates will be discussed again in the building section (p.162). In general, the microclimate in the intensive zones is the most important, which is why we spend the most energy on its improvement here.

Shelter and higher temperatures[6]

Good shelter reduces wind damage to leaves, flowers and fruit and can raise daytime air temperatures by 2°C or more, while soil temperatures can be increased by about 3°C. The higher temperature will make photosynthesis more efficient and reduce water evaporation. Overall, this will increase yields by 10–30%.[7]

If a windbreak is only installed in one direction, a north–south direction is preferable. This will stop the westerly wind, which is the prevailing wind direction in northwestern Europe, and the windbreak will not shade the sun much when it comes in from the south.

The windbreak should be as dense as possible. Shade-tolerant shrubs should be planted to prevent the windbreak from becoming leaky at the bottom, so that we avoid the wind from speeding up at ground level.

If the windbreak is to protect the home and living areas, it can also be a good idea to include evergreen shrubs and trees to provide shelter all year round.

Behind the windbreak, there is a quiet zone that is eight times the height of the windbreak. After this, there will be a zone with less wind and when we reach 16 times the height of the windbreak, there will be no effect. In an agroforestry farm with trees between the windbreaks, the wind will slide over these trees, extending

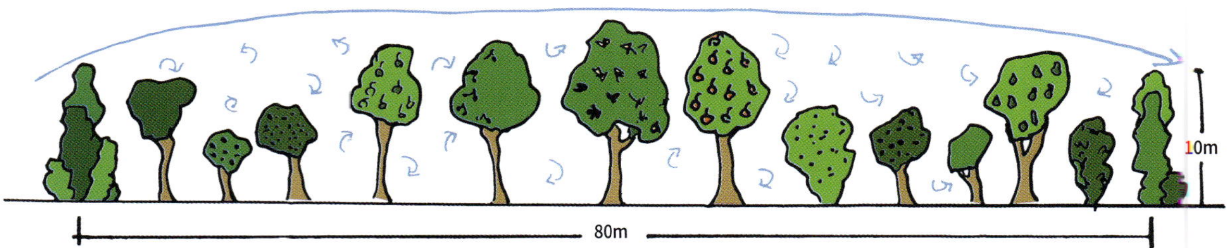

A dense windbreak creates a quiet zone that is eight times as long as the height of the windbreak. Here the air moves slowly and without direction. Ideally, the whole of zones 1 and 2 should be placed in the quiet zone.

Windbreaks must be dense at the bottom to prevent the wind from accelerating and hitting the production trees.

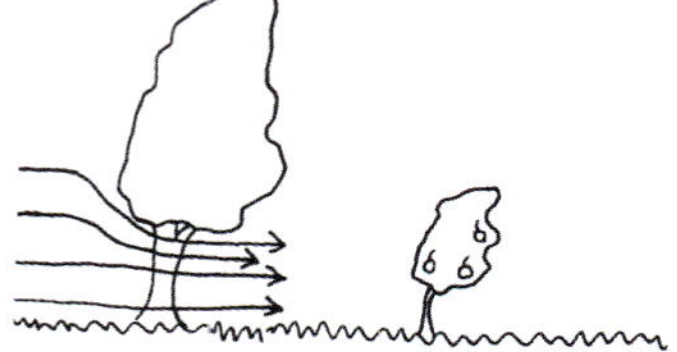

Turbulent winds will build up around the corners of a wall or a dense fence.

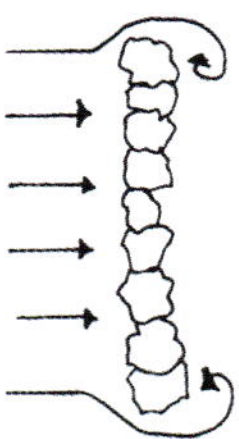

Windbreaks should be steep on the windward side so that the wind does not slide in an arc over them.

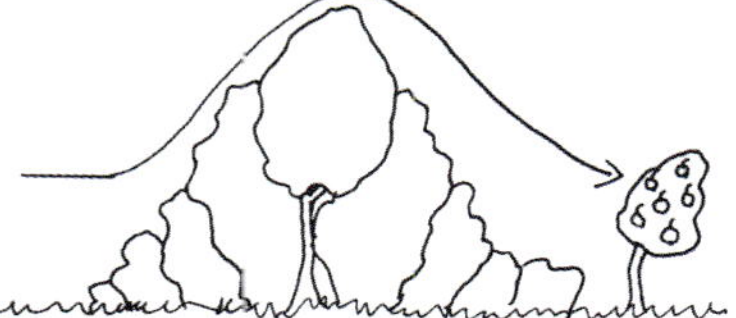

If the wind hits a wall or windbreak at an angle, wind will become concentrated along the windward side.

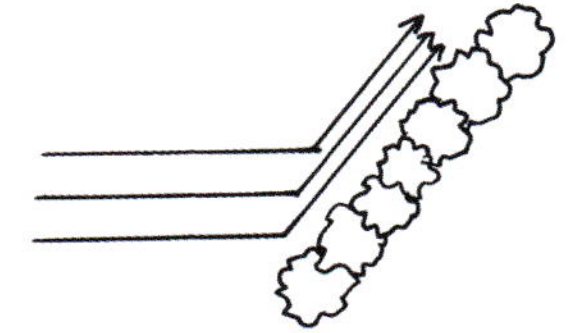

An optimal windbreak creates a quiet zone that reduces wind damage and raises the average temperature.

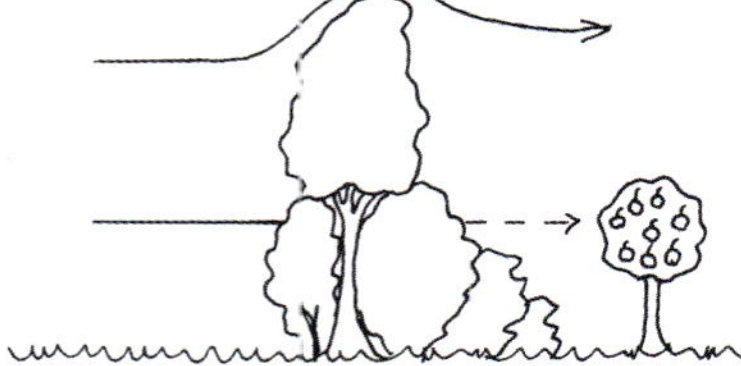

A narrow passage in an otherwise dense windbreak will speed up the wind. A short, dense fence behind the passage will stop the wind.

the quiet zone. It is usually best to position the windbreaks in the design so that all the intensively used areas (zones 0–2) are in the quiet zone.

Turbulence and concentrated wind

Windbreaks should be designed so that they reach all the way around that which requires protection, providing shelter from all sides and, at a landscape level, forming a cohesive network. This will shelter the entire landscape and avoid areas of turbulence and concentrated winds.

If it is not possible to create a continuous network, turbulence will occur around the corners at the end of a barrier such as a dense windbreak, hedge or wall. By planting more wind-permeable shrubs at the ends of the barriers, this turbulence can be broken.

Similarly, wind hitting these barriers at an angle will speed up and create an area of concentrated wind along the barrier. The same goes for a narrow passage in a barrier.

Other useful functions of windbreaks

The windbreak is not a waste of space, because here we can choose trees and shrubs with multiple functions, which

- Create habitats for insects that aid pollination and reduce pests – native species are especially good for insects.
- Collect nitrogen for the overall permaculture system if we plant nitrogen-fixing trees and shrubs.
- Produce berries and nuts, as long as we plant wind-resistant trees and shrubs.
- Produce leaf fodder.
- Produce winter feed for birds, e.g. native rose species.
- Produce chicken feed that the chickens can collect if they have access to the windbreak.

We can avoid the wind speeding up by letting a windbreak curve or twist and planting a short fence behind a passage. This is possible, for example, in market gardens, forest gardens and agroforestry systems, where cultivation and harvesting are carried out manually. However, if machinery is to be used or fencing is to be provided for animals, straight windbreaks are usually the most practical.

Frost pockets

Cold air sinks downwards in the landscape, and where it is slowed down by a dense hedge or where it reaches the lowest point in the landscape, a frost pocket will form, where the temperature is colder than elsewhere. In frost pockets, you should avoid planting heat-requiring crops, especially fruit and berries, as night frosts in late spring will damage the flowers, so no fruit will be set.

It is relevant to observe existing frost pockets while designing. The best way to do this is to see where there is frost in the landscape in the morning when the temperature is just around freezing. We should also be aware that frost pockets can be created when we build or plant hedges and fences.

Orientation of the landscape

The orientation of the landscape in relation to the cardinal directions affects

A frost pocket increases the risk of frost during flowering.

Sun exposure on south-facing slopes is far greater than on north-facing slopes.

the amount of sun exposure per square metre and thus the temperature of the landscape. On a south-facing slope, the sun's rays will fall densely, while they will be distributed over a larger area on a north-facing slope.

For every degree the landscape slopes to the south, the growing season is extended by two days, and conversely, for every degree the landscape slopes to the north, the growing season is shortened by two days.[8]

East-facing slopes will warm up quickly after a cold night. This is hard on cold-sensitive plants, which, if they have been frostbitten, benefit from a slow warming. South- and west-facing slopes are therefore better suited to these plants, also because photosynthesis is most efficient at a higher temperature, which will be in the afternoon.

Therefore, if possible, plant most species of fruit and berries as well as nut trees such as walnut and chestnut on south- and west-facing slopes. Robust species such as hazel and blackberry can be planted on east-facing slopes, while the north-facing slopes can be used for grass and forest.

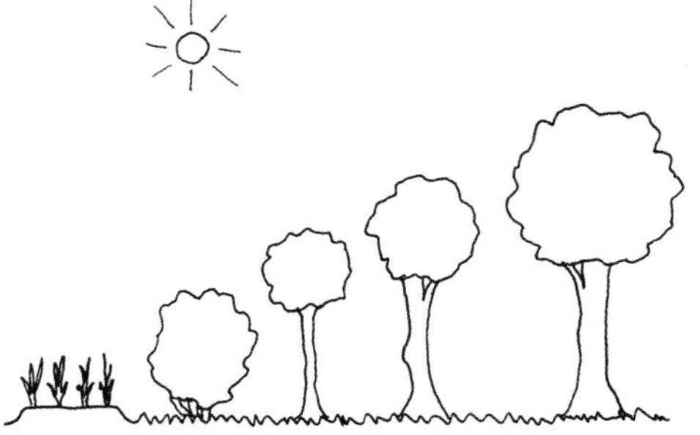

Planting designed as a south-facing forest edge.

A sun trap has dense, tall vegetation to the north and opens up to the south. The temperature and sun exposure is high and can be further increased by reflecting and storing heat in thermal mass.

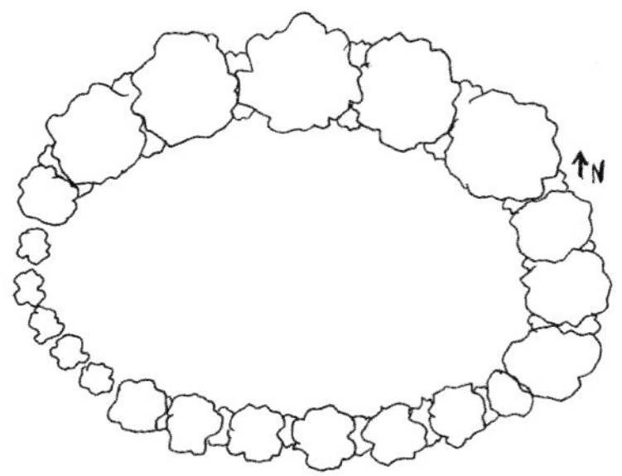

The clearing has the tallest trees to the north and the second tallest to the east because the morning sun is not as precious. The very lowest trees are planted to the west to bring in the all-important low-lying late-afternoon sun, and to the south, the second lowest. The clearing should preferably be designed elongated in an east–west direction to maximize sunlight.

Planting height profile

The placement of tall and low trees in the landscape can be used to create an area with greater sun exposure, much like a south-facing hill. For example, a food forest can be designed with the tallest trees to the north and gradually lower trees the further south you go. The canopy layer then becomes a long south-facing forest edge.

Sun traps and clearings

By creating sun traps and clearings, we create small areas where there is shelter and plenty of light so that the temperature is particularly high. Here we can place a home, a living area or grow heat-demanding low crops. The sun trap is adjacent to an open area to the south, whereas the clearing has trees all around. Tall trees are planted to the north. To the east, medium trees can be planted as the morning sun is not as precious as afternoon sun. To the south and west, there should be low trees and shrubs so that the afternoon sun and the low winter sun is allowed to enter.

It is important that the planting is dense at ground level so that cold air is not drawn in here. The air should be stagnant at the bottom, so that it can be warmed by the sun.

Reflection and thermal mass

The light intensity in a clearing or sun trap can be increased by creating reflection from a pond or, to some extent, from a plant with glossy leaves, e.g. autumn olive. The pond water is also a thermal mass that will retain heat from day to night, creating a higher night temperature in the immediate vicinity of the pond. Storing heat in thermal mass can also be done by placing stones in a clearing or making a clearing up a south- or west-facing wall.

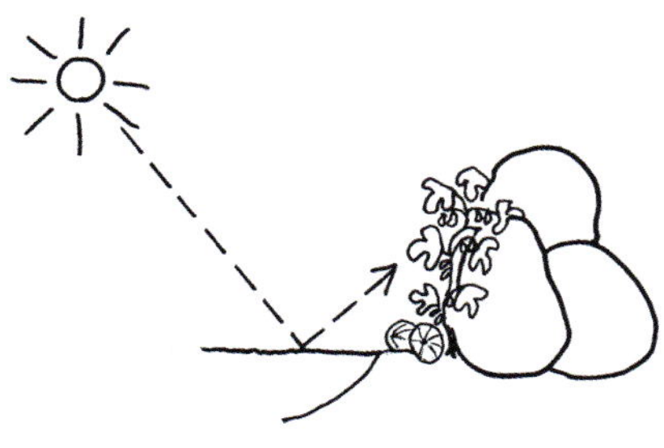

Light intensity can be increased by reflection from water or other shiny surfaces. Thermal masses like stone or water will store heat.

Forest gardens

What are forest gardens?

Forest gardens are long-term biologically sustainable systems for growing food and other products for a household or commercially.

In the UK these are called forest gardens, in Australia and America food forests, and in the tropics home gardens. If the forest-garden cultivation system is scaled up from a self-sufficiency garden to cultivation for sale, we have chosen to use the term food forest throughout, as it signals a larger area. But for the sake of simplicity, in this descriptive section we primarily use the term forest garden although the principles for a smaller forest garden and a larger food forest are the same.

In a forest garden, perennial vegetables and herbs, berry and nut bushes, fruit and nut trees, climbers and mushrooms are grown, which are all of direct and indirect benefit to people. Forest gardens are one of the examples of permaculture that most directly imitate nature's primeval forest with all its niches and layers and with all its lushness and resilience.

More precisely, it is the young part of a primeval forest that is imitated, where the crowns of the trees have not yet grown together blocking the light. This is so that nuts, fruit and berries get enough sunlight to ripen and so that light and air reach the ground cover layer. This partly means that clearings are established, creating more edges in the forest garden, but also that there must be a sufficiently large planting distance between the trees.

Forest gardens have been known for millennia in India and China, for example. The vision of forest gardens in temperate climates was formulated by Robert Hart in the 1980s. He established a small forest garden at his home in England. Martin Crawford, among others, has since done a great deal of development work on the cultivation method at the Agroforestry Research Trust in England. We have followed Martin Crawford's guidelines in our own work with forest gardens since 2007 and have found that they work very well in practice.

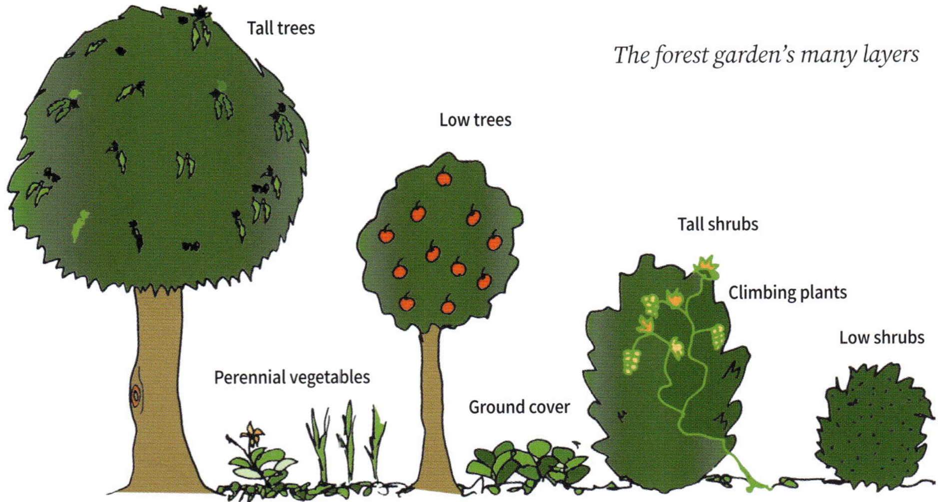

The forest garden's many layers

These guidelines have therefore been used in this introduction to forest gardening, combined with our own experiences. The spread of forest gardens has progressed rapidly in both Europe and North America in recent years.

Forest gardens are now a well-developed cultivation system within permaculture, and there is a lot of literature on the subject.

High resilience

The established forest garden is resistant to drought and only rarely needs watering. The forest garden is resistant to disease attacks and pests because the rich diversity of plants prevents diseases from spreading and also provides good habitats for beneficial predators.

Varied yield

The forest garden provides a varied yield that starts in early spring with perennial vegetables and peaks over the summer and fall with vegetables, berries, fruit and nuts, although some crops such as perennial kale can also be harvested in winter. In this way, the forest garden gives a high overall yield, but in contrast to a grain field, it is varied, distributed in different layers and has a longer harvest season.

When we have to establish an entire garden or forest with perennial plants, there will be a lot of formative work in the first years, but after that there will be only a little maintenance, e.g. pruning and weeding, so that harvesting work becomes the primary focus. The same forest garden can, with minimal maintenance such as replanting and weeding, continue to exist without having to be reestablished. Thus, the energy efficiency becomes high.

Energy efficiency

Where industrialized agriculture has a low energy efficiency, forest gardens/food forests are very efficient systems. This is because the inputs to industrial agriculture in the form of fossil fuels, stable buildings, machines, repetitive annual tillage, sowing and watering as well as the production of fertilizers and pesticides, etc. are huge.

The inputs to forest gardens are minimal because we let nature work for us.

CECIL RYE-OLSEN

Forest garden in fall colors. Inspiratoriet, Denmark.

SKYE JIN

Forest garden with all layers and great diversity in the ground cover layer provides self-sufficiency with greens, berries and fruit for a long season. Skye Jin, Denmark.

Fall is sumptuous in the food forest, providing berries, fruit, nuts, leafy greens, herbs, mushrooms, eggs, etc. for a healthy, local luxury diet that counteracts climate change. Reforest Farm, Denmark.

Winter forest garden salad with frozen silverberries, rich in omega 3 fatty acids and vitamins.

Forest garden salad with greens, berries, flowers and nuts, fried sprouts, vegan pesto and tzatziki with orpine leaves.

October food from the food forest with roasted chestnuts, which are rich in omega 3 fatty acids, and forest garden salad. A long season of fresh produce.

So even if the yields measured in calories will in some cases be higher in industrial agriculture than in a food forest, the energy efficiency, where we divide energy used by energy in the yield, is far lower.

Forest garden food

Food is very varied with nuts, fruit, berries, seeds, flowers, leafy greens and mushrooms. It is a healthy, varied diet that is fresh for a large part of the year.

Perennial vegetables have a larger root system than annual vegetables, and this larger root system means that the plants tend to have a higher content of minerals, vitamins and proteins.[9] A test of 14 different perennial vegetables has shown that the species mulberry (the leaves of the tree), patience dock, Good King Henry and nettle in particular were very concentrated in their nutritional content.[10]

With new plants, of course, comes a new diet and new cooking. A forest garden diet is locally produced, it counteracts climate change and is anything but boring.

Canopy layer and shrubs

It is from the canopy layer, which receives most of the direct sunlight, that the greatest yield in the forest garden is obtained.

Common trees such as apple, pear, cherry, plum and hazel are obvious to plant, but there are also a number of other trees that are less well-known that can help to provide a nice variation in the diet. These include paw paw, which bears large fruits with a taste like a mixture of mango and banana; mulberry, which produces large mild 'lettuce' leaves and raspberry-like fruits; Sichuan pepper with strongly flavored seeds; Chinese cedar with spicy leaves and sweet tasty flowers; or exotics such as peach, apricot and date plum, to name just a few.

Everyone recognizes a peach, but to be able to grow varieties in temperate climates is a recent change. Here they are planted in a row on the north side of a clearing where they benefit from a warm microclimate. Reforest Farm, Denmark.

Also in the shrub layer, well-known shrubs such as gooseberries, redcurrants, blackcurrants, blackberries and raspberries will be good choices. Of less common species, choose, for example, honeyberries with blueberry-like berries in the spring; aronia with aromatic berries; or the species of nitrogen-fixing silverberries that produce berries, e.g. autumn olive or goumi.

Design of the canopy layer

It is important to make a design for trees and larger shrubs before the forest garden is planted so that we get a good zone design, the right planting distance and beneficial relationships.

A simple design method is to make a map of the landscape with observations of shelter, soil type, humidity, frost pockets, etc. Movable pieces are made out of cardboard to show the fully grown crowns of the trees and shrubs.

BENJAMIN BRO-JØRGENSEN

Design for a larger and simplified commercial food forest, primarily with aronia, hazel and nitrogen-fixing autumn olive. In addition to fresh products, jam is also sold. The food forest is on a southwest facing slope, and the shrubs are planted in rows along the contour lines.

BENJAMIN BRO-JØRGENSEN

Late-summer in the same established food forest. Permakulturplanter.no in Tingvoll 800km north of Oslo in Norway.

The pieces can then be moved around on the map and adjusted. This can also be done on a computer. Take your time on the design; these are trees that must stand and produce for 30–200 years, and it is far easier to move a piece of cardboard than to move an established tree.

The most common mistake made in forest gardens in temperate climates is that the small young trees are planted far too close together, with the result that as the trees and shrubs grow, the forest garden turns into a dark and impassable thicket with no light reaching the ground cover layer. In this way the forest garden also becomes difficult to harvest, and the microclimate becomes humid, so fungal diseases can easily spread.

As a rule of thumb, the planting distance should be at the fully grown crown

Tree Planting

When we plant production trees and shrubs, it pays to make the most of the planting to get the best establishment and growth.

The best thing is if we can buy the trees as relatively small bare-root trees. Fruit trees from a nursery will often have roots that twist in the pot. Here the roots must be straightened out. It is better for some of the roots to break than for them to twist around in the planting hole. Take care that the roots do not dry out before adding soil.

The planting hole is dug to the same depth as the root ball and so wide that even long roots can be straightened out. Only with very compacted soil is the soil loosened deeper, as the tree must have solid ground to stand on. The planting hole is dug wide enough that the roots can point away from the tree without being folded into the hole.

Experience has shown that mixing the soil used to refill the hole with shells makes it less attractive for voles to dig around the roots.

The tree is placed in the hole, the soil is filled back in, and it is lightly tamped around the roots. Soil is only filled up to the point on the tree where the root becomes a trunk, because if soil gets on the bark it can cause rot.

Finally, water well with at least 10 liters of water to reduce air pockets in the soil.

It can be practical to make the soil surface around the finished tree slightly bowl-shaped, then it is easier to water the tree afterwards. During a drought, a newly planted fruit tree should, as a rule of thumb, have 10 liters of water once a week.

Protection of newly planted trees

Newly planted trees must be protected from grass and damage from animals. Grass is a tough competitor for water and nutrients, so the tree will grow significantly slower and sometimes die completely if it grows in grass. Grass is most easily kept away by placing a piece of cardboard, woven ground cover fabric or black plastic on the ground around the tree for the first three years. The ground cover must keep the grass away for a radius of 50cm from the tree. Strongly growing perennial vegetables can also give too much competition to a newly planted tree and must be kept at a distance. It may be necessary to remove the ground cover from the trees in winter, as mice that gnaw the bark of the trees can hide under it. Newly planted trees should be protected against damage from hares, mice, deer and grazing livestock. This can be done both individually with plant tubes or plant spirals, and by making game fences around the planted area.

Sichuan pepper (Zanthoxylum schinifolium) is a small tree 3–4 meters tall and wide that yields an abundance of strong-tasting seeds that can be put in a pepper grinder. Agroforestry Research Trust, England.

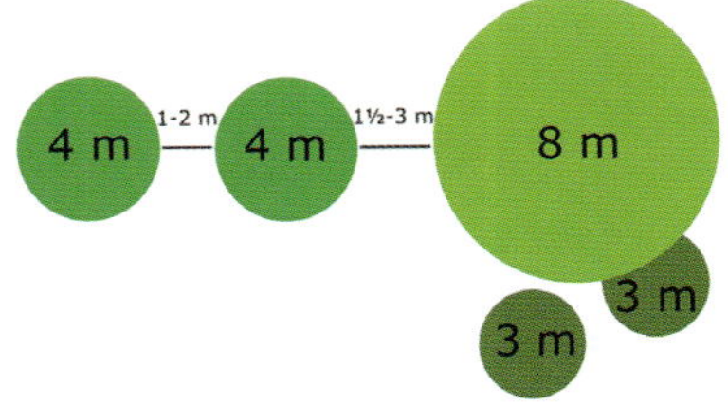

The planting distance should be based on the size of the fully grown crown, plus 25–50%. If the crowns have different widths, an average is taken. If a small tree or shrub is planted on the south side of a tall tree, it can be placed partially under the crown of the tall tree and still get the light it needs.

plus 25–50%. If the crowns have different widths, an average is taken.

There are quite a few exceptions to this rule. If a small tree or shrub is planted on the south side of a tall tree, it can be placed partially under the crown of the tall tree and still get the light it needs. Other tall trees such as robinia are so light-open in the crown that shade-tolerant small trees such as a coppiced lime tree, from which the leaves are harvested, and shade-tolerant tall shrubs such as plum yew or bamboo, with edible shoots, can be planted right under the tree.

When the design is made, pollination must also be considered as an important beneficial relationship. Trees such as apple, pear, cherry and plum are cross-pollinating, so two different varieties that flower at the same time must be planted near each other. Other species are dioecious, meaning there must be a male plant for the female plants to be pollinated. This is the case for sea buckthorn and gooseberry kiwi. We need to research the pollination conditions for the various species in the forest garden and take this into account in the design.

Mulberry (Morus alba × rubra) grows to 10 meters tall but is pruned at harvest height. It produces sweet, stoneless fruits in summer. The young leaf shoots, which the tree sets partly in the spring and again in September, are a delicate leafy vegetable, e.g. fried. Chickenforest at Reforest Farm, Denmark.

Date plum (Diospyros spp.) can bear rich, exotic-tasting fruits that are harvested in November and ripen on the windowsill. The tree here is a cross between common persimmon and American persimmon. Reforest Farm, Denmark.

Chinese cedar (Toona sinensis), from which the seasoned leaves are used e.g. in stews, is very healthy. Reforest Farm, Denmark.

Date plums ripening on the windowsill.

The fruits of this white mulberry ripen over a long season.

Paw paw (Asimina triloba) produces creamy fruits with a taste and texture similar to banana and mango. The fruits ripen in September to October. Reforest Farm, Denmark.

The large-fruited hawthorn (Crataegus ellwangeriana) produces sweet fruits that can be eaten raw and used for jam, ketchup or wine. Reforest Farm, Denmark.

Bladder nut (Staphylea pinnata) becomes a shrub 5 meters tall. The nuts sit in clusters. The flowerbuds can be eaten in the spring, and especially delicious are the unripe crunchy nuts, which taste fresh and sweet like peas in the middle of summer. Reforest Farm, Denmark.

Shelter and microclimate

Before the canopy layer is planted, it is important that shelter is established so that a good microclimate is achieved. Many fruit trees are very sensitive to wind. Fruit trees established in windy conditions will often produce a lower yield than trees established in a sheltered microclimate.

It is obvious to place the forest garden adjacent to an existing forest or shelter-belt, as it will not only provide shelter, but will also help to ensure that a rich insect life and a forest floor environment will emerge in the forest garden more quickly, which together will promote growth, yield and a robust cultivation system.

Ground cover and perennial vegetables

Although the greatest yield is obtained from the canopy layer, which receives most of the direct sunlight, the ground cover layer is also important. It has a wide range of functions that support the entire cultivation system while at the same time providing a yield from perennial vegetables.

Early harvest – In the spring, the forest garden will come into leaf from the ground upwards; ground cover, then shrubs and finally the trees. Thus, perennial vegetables can be harvested from early spring, a time when annual vegetables have not yet been sown. Fresh greens are welcome after a long winter. Other perennial vegetables come later, and we can harvest them until the winter.

Diversity – In the forest garden's canopy and shrub layer we may have 10–30 different plant species, while in the ground cover layer we quickly come up with a hundred or perhaps several hundred different plant species and varieties. This is where we find the greatest diversity,

ANDREAS JONSSON

The lower layers emerge first and use the early-spring sun. The forest garden in Holma, Sweden.

which is important to ensure the presence of pollinating insects and create resistance to pests and diseases.

The ground is kept covered – The ground cover layer keeps the ground covered all year round, either by the plants being evergreen or via the dead stems of the plants that lie on the ground. By having a dense plant cover that takes care of itself, we minimize weeding. When we also choose plants that only grow to the desired height, we don't have to mow it like we do with grass.

Functional plants – In the ground cover layer, we can incorporate accumulator plants, which with their deep roots retrieve minerals from deeper soil layers, e.g. potassium and phosphorus, which are primary nutrients, but also a large number of others. The minerals will then be in the leaves of the plants, and when they break down they will benefit other plants in the forest garden. Comfrey is the accumulator plant that picks up the most minerals, and here we can use the sterile variety Bocking 14, which doesn't spread by self-seeding. A large number of the perennial vegetables are also accumulator plants, such as all the sorrels, rhubarb, sweet cicely and perennial fennel.

Many of the plants in the ground cover will be attractive to beneficial animals and aromatic plants, such as mint, lemon balm and oregano. They release essential oils that have an anti-fungal and anti-bacterial effect. This is believed to help distract pests and protect nearby plants from fungal and bacterial diseases.

Design of the ground cover layer

While the most frequent design error with the canopy layer is too small a planting distance, too large a planting distance is the most frequent error when establishing the ground cover. The ground cover plants must be planted just slightly closer together than the distance that the fully grown plant will reach, and in staggered rows, so that the plants cover the ground completely during one growing season. It requires a lot of plants and is immediately a lot of work, but in this way, the new ground cover only has to be weeded in year one, and after that it is mainly just harvesting. Below is a little more about the design and establishment of the ground cover.

Clump-forming and spreading plants

Perennials can generally be divided into two types of plants.

Spreading – These plants spread by sending out above-ground stems that take root and become a new plant, like strawberries. Raspberries send out underground roots that shoot up as a new plant. Other spreading plants include ostrich fern with edible shoots in the spring (must be boiled for 15 minutes), bistort with mild-tasting leaves for the salad bowl or to be used like spinach, and wild strawberries, as well as ground ivy, barren strawberries and sweet violets.

Clump-forming – These plants grow to form a clump or a small woody shrub and only spread with seeds. Many of the perennial vegetables are found in this category. This applies to daylilies with large edible flowers; orpine with crisp 'cucumber-like' leaves; sweet cicely with edible flowers and leaves and seeds that taste mildly aniseed; Good King Henry with 'spinach leaves', protein-rich young flowerbuds and quinoa-like seeds; udo with crisp stems; scorzonera with mild tasting leaves; patience dock with giant leaves; a wide variety of perennial bulbs and many more.

STEPHEN BARSTOW

The spring shoots from many species of hosta can be eaten raw or cooked. Stephen Barstow, Norway.

Orpine can be eaten from early spring until it blooms in early September. Here it is in a polyculture with daylilies and Turkish rocket. Reforest Farm, Denmark

Russian comfrey 'Bocking 14' is the best accumulator plant and does not spread by seed.

Patience dock (Rumex patientia) has huge, mild-tasting leaves at the end of April. It withers down in summer to shoot again in September. Reforest Farm, Denmark.

Turkish rocket (Bunias orientalis) produces broccoli-like flower buds in early summer. Later and well into fall, the large green leaves can be eaten. Since it is not in the cabbage family, it is not attacked by caterpillars. Reforest Farm, Denmark. Denmark.

Good King Henry (Blitum bonus-henricus) produces soft spinach-like leaves and forms a good ground cover. It is also a perennial grain and quinoa-like seeds can be harvested from the stalk. Reforest Farm, Denmark.

Good King Henry unopened flower buds are incredibly healthy and full of protein. They are sautéed in a pan and have a wonderful nutty flavor. Reforest Farm, Denmark.

Sweet cicely (Myrrhis odorata) tastes like aniseed, and the leaves, flowers and seeds are all edible. It is a good bee plant and accumulator plant. In the foreground of the photo they are seen with white flowers. Reforest Farm, Denmark.

Remove perennial weeds

To ensure that the forest garden does not end up being overgrown with perennial weeds, such as nettles, it is important to get rid of all perennial weeds before the ground cover layer is planted. This is done by preventing the weeds from getting light for a whole year.

In a larger food forest, it is most effective to lay a water-permeable woven ground cover fabric on the ground for a year, so that there is a completely weed-free soil when planting. You can make a few holes in the fabric and plant different kinds of pumpkins, which grow across the fabric. The ground cover plants must be planted closely as soon as the fabric is removed so that they come up and cover the ground before the weeds. Another way to establish new beds is the 'lasagna method'.

Lasagna gardening method

Lasagna gardening is a method for establishing new beds, where we simultaneously shade away the perennial weeds and add large quantities of compost to the soil. With the lasagna method, we also use the first year for plant growth and avoid digging the soil. The method can be used both for the establishment of perennial ground cover plants in a forest garden or for the establishment of a raised-bed garden with annual vegetables.

If the soil is compacted, loosen it with a broadfork, wiggling it slightly from side to side. A carpet made of biodegradable materials or old bed linen is then laid to shade away the existing vegetation. Compost is placed on top, possibly mixed with weed-free topsoil, and finally a layer of straw to reduce drying out of the compost. Plants are immediately planted by scraping the compost aside and placing two handfuls of soil in which to plant. The roots themselves find their way through the fabric as it breaks down.

Cardboard shades more effectively than fabric, but since cardboard often contains PFAS, it may be a good idea to limit the use of this. It is easier for plant roots to establish themselves through woven carpets and linen.

However, the lasagna method is relatively labor-intensive and requires large amounts of compost. It is most realistic to use this method in a smaller forest garden. On larger areas, shading away with plastic ground cover fabric and subsequent planting will be more affordable.

In summer, Scorzonera (Scorzonera hispanica) stems and flower buds are delicious sautéed or in a wok dish, and the root can be eaten in winter. Reforest Farm, Denmark.

ESBEN SCHULTZ

Daylily (Hemerocallis spp.) is grown in Chinese forest gardens, where the large, crisp flower bud is used in wok dishes, or the flower can be dried and used in stews. The fresh flower is also used in salads. It forms dense ground cover. Reforest Farm, Denmark.

STEPHEN BARSTOW

Udo (Aralia cordata) comes from Japan and is probably the fastest growing and largest perennial vegetable. It shoots 2–3 meters from the ground every year. The stem is peeled before eating and has a crisp, lemony flavor. Norway.

Giant Solomon's seal (Polygonatum commutatum) shoots up at the end of April, and the stems have a crisp and pea-like taste. The flower bud is broken off as it has a bitter taste. Here it is seen in a ground cover of ground ivy. Reforest Farm, Denmark.

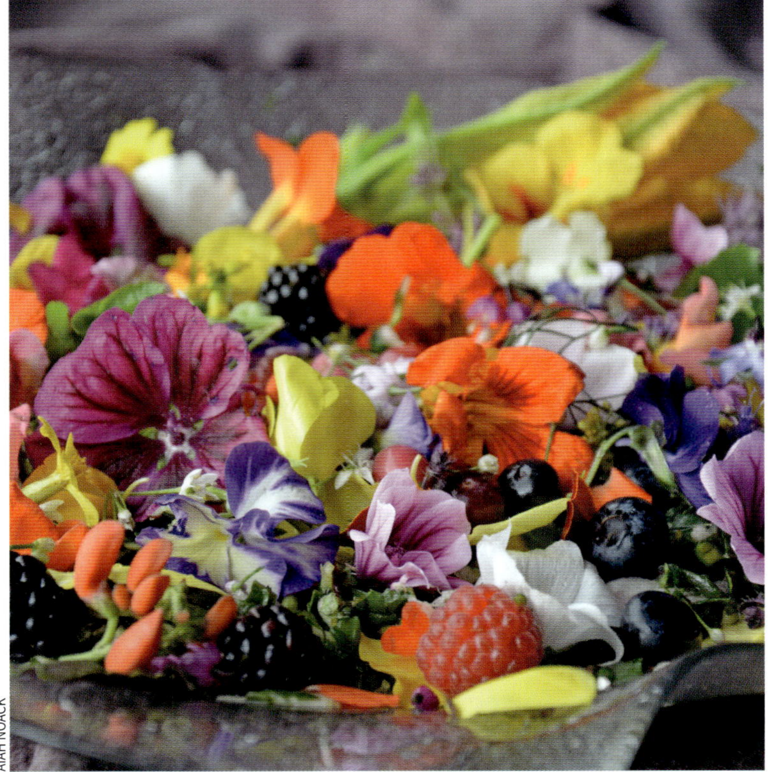

AIAH NOACK

Beautiful salad of perennial vegetables, flowers and berries. Naturplanteskolen (the Nature Nursery) in Denmark.

Perennial buckwheat (Fagopyrum cymosum) is a leafy vegetable that grows to 2 meters tall. It comes in late spring and continues to make new fresh leaves along the stem until late autumn. Therefore, it is one of the most productive perennial vegetables. The leaves are mild and can be eaten raw in salads, hummus and pesto. It spreads with underground root runners and covers the ground completely. Reforest Farm, Denmark.

AIAH NOACK

Perennial cabbage is propagated by cuttings because, as a perennial, it does not flower. Stick the stem with its leaf on in the soil in a greenhouse in February and they will root easily. Naturplanteskolen, Denmark.

KRISTINE FENGER

Perennial cabbage is a very productive winter vegetable. Here it grows under apricots in fall at Kristine Fenger, Blomst&bønne, Denmark.

Bistort (Persicaria bistorta) is a productive perennial green with a long harvest season. It functions as a ground cover plant that spreads by underground shoots. It is a native plant, and when the beautiful light purple flowers pop out in May and June, it attracts many insects. Reforest Farm, Denmark.

BENJAMIN BRO-JØRGENSEN

Long rows of perennial vegetables established through a compostable weed cloth at Permakulturplanter.no. From here, salad of perennial vegetables are harvested, which Benjamin Bro-Jørgensen takes to the local café. Norway.

The ground has been covered with woven ground cover fabric for a year, and now cuttings of orpine (Sedum telephium) are put into the ground. Orpine is a clumping perennial whose leaves are juicy and mild to eat from early spring until September. Reforest Farm, Denmark.

Ostrich fern (Matteuccia struthiopteris) is a perennial vegetable that spreads by root shoots. It can be established with the expanding edge method and grown in a polyculture with ramsons (Allium ursinum). The fresh shoots must be boiled for 15 minutes and taste like artichokes. Reforest Farm, Denmark.

ESBEN SCHULTZ

Scorzonera can be harvested throughout the year, with all parts edible. In the spring and again in September, the green leaves are eaten as a salad. Reforest Farm, Denmark.

Expanding edge

The vigorous spreading ground cover plants can be established using the 'expanding edge' technique. A number of these are planted and ground cover fabric is laid on both sides. It shades out the weeds, and at the same time the spreading ground cover plant shoot their roots under the plastic or their stems over it. The weed cloth can then be continuously pulled outwards and the creeping ground cover will quickly occupy the weed-free soil. In this way, the work of planting the many ground cover plants is saved, and we do not need so many plants to get started.

STEPHEN BARSTOW

Caucasian spinach (Hablitzia tamnoides) is a perennial that comes up in early spring and grows here on sticks into a tree. The plant gets stronger every year. The leaves have a fresh, mild taste. Norway.

Climbers

Climbers can be grown up trees and shrubs, so you avoid having to build a frame. They give the forest garden a lush and luxurious look and help to further optimize the yield. In practice, climbers find it difficult to cope when planted right next to a tree trunk, where there is not much light and where there is a lot of competition for water and nutrients. Therefore, plants such as grapes, kiwi, caucasian spinach, etc. are planted a meter or more from the trunk of a tree and a stick is placed for it to use to climb the tree. Choose a tree that does not grow too tall so that it is possible to harvest fruit or leaves from the climber, and a tree that does not require full sun, as the creeper will provide shade. It can, for example, be a lime tree that is kept at a manageable height for harvesting the leaves or the shade-tolerant Ebbing's silverberry, *Elaeagnus* x *ebbingei*.

Polycultures

Trees, shrubs, ground cover plants and climbers are grown together in polycultures, where the plants support each other. Thus, there should preferably be nitrogen-fixing plants, plants with deep roots, plants with shallow roots, plants that cover the ground and plants for insects in every bed in the forest garden.

Examples of good polycultures can be ramsons, which come very early in the spring, planted together with ostrich fern, which will cover the ground when the ramsons wither in the summer. In this mixture could be a strong accumulator plant such as comfrey or rhubarb with deep roots. In addition, plant a nitrogen-fixing large shrub like autumn olive on the south side and a tall fruit tree to the north. Many combinations are possible.

In a larger food forest, we recommend that each polyculture in the ground cover layer contains no more than three species and that each bed covers a reasonably large area so that the food forest becomes manageable to look after and harvest.

Paths

In a cultivation system without tillage, it is important that the soil is not compacted by walking on it, and therefore permanent paths are established.

Depending on the scale, we can build tiled paths, paths with wood chips or mowed grass paths and possibly combine them with stepping stones in the beds. They can be designed in such a way that we can completely avoid stepping in the beds.

In a larger food forest, the realistic thing would be a network of main paths, and that we occasionally step outside the main paths and into the ground cover to harvest.

The main paths are advantageously placed between the fully grown crowns of the trees so that we can harvest from them and so that over time it does not become necessary to cut branches from the trees to keep the paths clear.

Animals in forest gardens

It is also possible to keep animals in a forest garden as an extra 'layer' and to obtain an extra yield. It will be easiest to

STEFFEN MEYER

At Gammelgård in Sweden, Steffen and Christina Meyer have established a forest garden for self-sufficiency with high diversity. The paths are maintained with wood chips.

incorporate poultry, as they do not compress the soil and do not destroy the trees. Ponds with fish are also an option. See the section on animal husbandry, p.68.

Mushrooms in the forest garden

Over time, by pruning and felling, it will be possible to harvest wood from a forest garden, and innoculate the logs and the stumps with mushroom spores. Logs are usually inoculated with purchased mushroom spores of decomposer mushrooms, such as oyster mushroom and shiitake. Felled wood can be used after roughly six weeks, once the tree's natural fungicides have been released and died off. Logs need to be used relatively fresh like this, to avoid wild fungus taking over. Spring is best for inoculation as there are fewer fungal spores in the air, and frosts need to be avoided. The inoculated wood is stored in a shady place where they can easily be monitored.

Many variants

Forest gardens can be small, intensive and varied: forest gardens in a backyard or a villa garden (zone 1), possibly with just a single tree in polyculture with shrubs and perennial vegetables. But as mentioned, food forests can also be established on a larger scale with the aim of being able to sell a large selection of products from zone 2, or simplified to only contain selected species in zone 3. Here, forest gardening will have a smooth transition to agroforestry/permaculture farming.

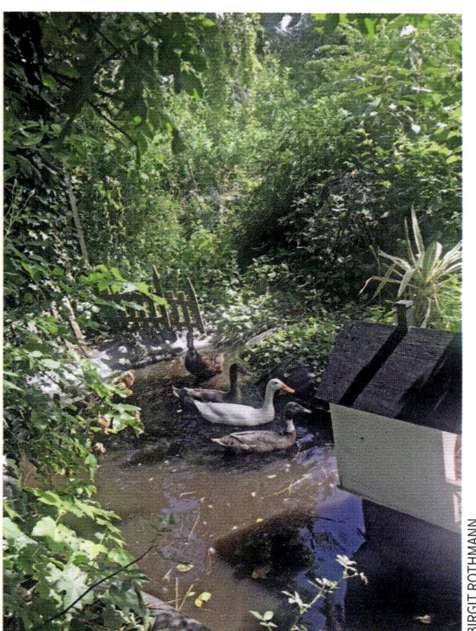

BIRGIT ROTHMANN

Ducks in the forest garden on Kærbakken. Denmark.

SIMON NYRUP MADSEN

In Svanholm Forest in Denmark, mushrooms are grown as an extra benefit.

Permaculture farming and animal husbandry

In recent years, there has been a positive development toward permaculture in parts of existing industrial agriculture. Therefore, there is a need for a clarification of the concept.

Permaculture farming

Permaculture farming can include several types of cultivation systems, with or without trees and animals. The more permaculture principles that are included, the further we get toward a resilient system. The regeneration of natural resources and the mitigation of climate change also increase, the closer we come to using nature as a model and thus incorporating the most permaculture principles. If only a few permaculture principles are used, the overall system will often not be regenerative and will not counteract climate change.

Examples of permaculture farming are alley cropping, food forests, silvopastures with production trees or fodder trees and animals, cultivation of annual vegetables in no-dig beds, also called market gardening, etc. Permaculture farming is always based on a thorough design. Permaculture is also defined as being regenerative in a holistic sense, where all four natural resources are regenerated.

Agroforestry

Agroforestry covers all agricultural systems that involve trees; food producing, fodder trees, trees for timber and windbreak trees. Most permaculture farms are based on trees, and the more trees, the greater the carbon storage. Larger industrial farms around the world have also begun to include rows of trees in the field, under which annuals are cultivated with plowing or animals are grazed. There is a gradual transition between what is defined as large-scale agroforestry and what is defined as permaculture farming. An agroforestry farm with enormous fields, where rows of fruit trees on weak rootstocks are planted 40 meters apart and annuals are grown in a system with plowing between the rows cannot be defined as permaculture. In a permaculture version of agroforestry, the trees will usually be what shapes the design and therefore occupy much more space.

Market gardening

The term market gardening actually refers to the fact that there are sales from an intensive, expanded garden. In market gardening, annual crops are often grown in no-dig beds, a cultivation system that is also used in a permaculture context, as it is the way in which we can produce annual vegetables in line with most of the permaculture principles.

Regenerative agriculture

One might get the impression that this is the same as permaculture farming, which is also about regenerating. There is an overlap between the principles of

Permaculture design for the farm Hjortespringgård in Denmark that includes a market garden with annual and perennial vegetables, a food forest, a mobile chicken coop, an apple orchard, a permanent pasture where sheep roam, a newly dug pond for biodiversity and a zone 5 forest. Thus, it is a good example of how several systems are combined in a permaculture design.

regenerative agriculture and permaculture, but there are also important differences. The concept of regenerative agriculture is used by many different actors in agriculture and encompasses very different agricultural systems. From small-scale market gardens to large-scale farms that sell their products to international corporations. The key is to reduce soil treatment by, for example, rotavating or harrowing the fields rather than plowing, or to completely switch to no-till farming and ensure that the soil is always kept covered, with living or dead plants. Farm animals must be grazed and integrated into cultivation systems, for example with ruminants in holistic grazing. Work is done on crop rotation and on increasing biodiversity. Unlike permaculture, there is not necessarily a focus on thorough design prior to the start-up of regenerative agriculture. The term regenerative agriculture covers both systems that can be called permaculture and those that are so far removed from the natural ecosystem that they cannot. While regenerative agriculture in some cases is mostly focused on regenerating the topsoil, permaculture is always regenerative in a holistic sense.

The spread of permaculture farming

Permaculture is a global movement. The permaculture principles are the same everywhere, but the cultivation systems are adapted to climate and culture.

There is a tendency for permaculture farming to spread faster in areas where climate change causes a lot of instability in the weather and the seasons so that the usual cultivation of annual crops becomes too risky. Cultivation with polycultures of perennial crops is, as described, more resistant to, for example, drought or extreme rainfall.

There is also a tendency for permaculture farming to spread faster in countries where small, family-owned farms are still in existence. This is because it is more difficult financially to convert a huge industrial farm to permaculture farming, since loans from banks have already been invested in land, large machines and buildings.

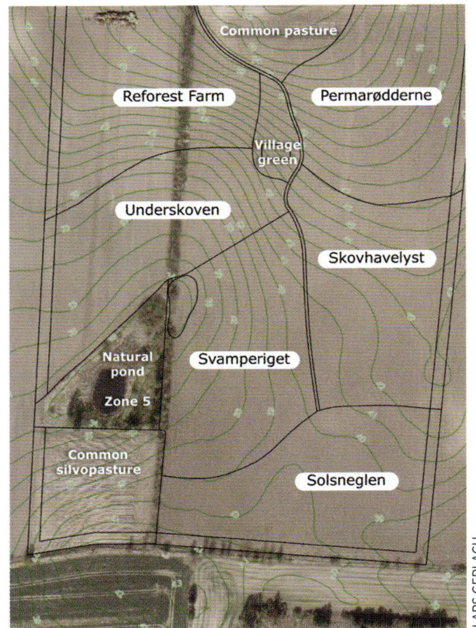

The eco-hamlet Skovvirke at Svanholm organic farm in Denmark has been granted planning permission to create six new mini farms based on permaculture. The farms have 1.5 hectares of land each and modest residential houses and farm buildings. The permit has been granted under special conditions inspired by the One Planet Development policy in Wales. The conditions ensure that permaculture farming is practiced and that the residents live with a sustainable ecological footprint. The project is an exception to the trend; since the 1950s farms in Denmark and many other countries have been consolidated.

Structural change to small scale

In countries that currently have industrialized agriculture, the transition to permaculture farming also entails a restructuring so that farms become smaller, according to the principle of small scale. There are different proposals for how this structural change can be implemented. Shared ownership, where several farmers take over a farm with existing buildings and carry out collaborative projects; new small farms with new buildings, where existing large farms are broken up; and cooperative and consumer-owned farms are just some examples of structures that make small intensive agriculture possible.[11,12,13,14,15] The organization La Via Campesina[16] works politically to make small farming possible and fights for the rights of small farmers. The economics and structures surrounding agricultural transformation will be described further in Chapter 6.

One consequence of the transition to permaculture farming will be that more people will work directly in agriculture than the few percent who do so today in industrialized countries, but there will still be significantly fewer than before the industrialization of agriculture.

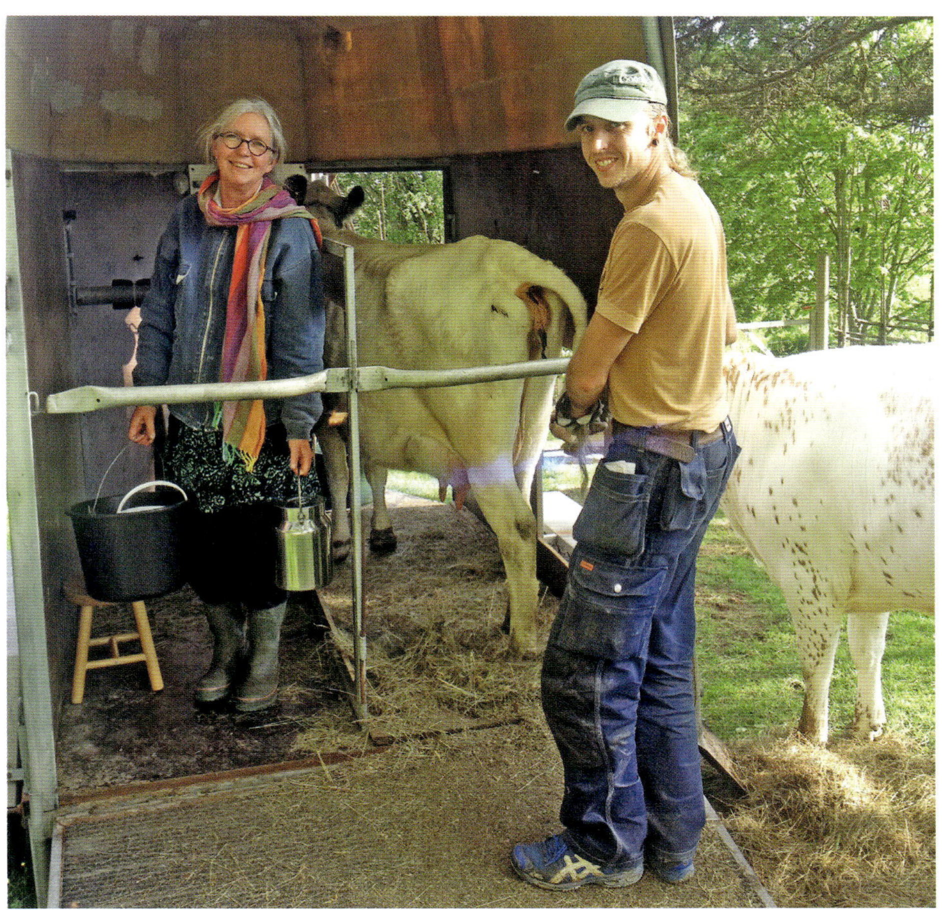

At Rikkentorp in Sweden they milk cows by hand and sell the milk locally as a political statement – because at the same time, Joel Holmdahl is active in the organization NOrdBruk, the Swedish member organization of La Via Campesina, which works for better political conditions for small farms.

When Cuba lost its supply of fossil fuels after the collapse of the Soviet Union, 15–25% of the population had to work in agriculture, including using permaculture techniques.[17]

But this was in a crisis situation, so we must assume that the proportion will be significantly lower if the conversion is well planned and takes place in a time of greater surplus. Permaculture farming does not have to be without machinery, but it makes us less dependent on large machines with a lot of horsepower. We must also balance the use of machinery against the fact that we have to produce food with a much lower ecological footprint than in the industrialized food system.

Extra work during the transition period is a necessary investment in our own future and the Earth's. It is a prerequisite for preventing biodiversity loss, regenerating natural resources and counteracting climate change.

Animals and permaculture farming

Permaculture farming will often work with animals as part of the cultivation systems. To design a permaculture system with animals, permaculture principles are used, i.e. *stacking, diversity, small scale, beneficial relationships,* etc. In particular, it is central that the animals have *multiple functions,* since an animal that is only kept for meat production is not an energy-efficient way to make food – ten energy units in feed become approximately one energy unit in meat.

For this good reason, many who work with permaculture are also vegetarians or vegans, while others have a reduced meat consumption. In any case, it is difficult to make large meat consumption consistent with a permacultural lifestyle.

But animals are also part of the natural ecosystem and thus our model for permaculture. In nature, however, there are far fewer animals per area than in optimized industrial agriculture, and thus there will also be fewer animals in permaculture farming.

If we design our permaculture systems so that the animals have multiple functions – where we let them harvest their own feed, where they create an extra 'layer' under trees and where we make use of their heat and/or muscle power – the calculation becomes completely different, and the animals can in some cases contribute positively overall to the permaculture farm.

If we first look at animals such as chickens or pigs in industry as a contrast, they are kept indoors in giant barns with poor living conditions. They are fed grain from agriculture with plowing, which degrades the soil and has a high climate impact from the use of fossil fuels. They are also fed soya imported from the world's poor countries, where the proteins could have fed far more people and where forests have often been cleared to make fields, resulting in a major loss of biodiversity. The animals' only real function is to produce meat. Such production is destructive and energy inefficient.

Permaculture farming with animal husbandry starts in a completely different place by using nature as a model. It is knowledge about the primeval forest and the natural living conditions and behavior of the animal in question that must be combined so that beneficial relationships are achieved. Self-feeding systems, where the animals harvest their own food, and systems where the animals have multiple functions, are central to permaculture animal husbandry.

One advantage of some of the smaller animals, such as chickens, ducks, geese, pigs, sheep and goats, is that when they have young in the spring, they will have to be slaughtered in the fall. This means that we only need to feed them during the summer months, when nature in a temperate climate is highly productive. This allows us to rely as much as possible on self-feeding systems and save the work and energy consumption of harvesting and storing winter feed and finally feeding it to the animals. We can produce meat that follows the course of the year. And as we will see in the coming sections, the animals' feed harvest will often be a useful function in the farming system.

The chicken as an example of an input–output analysis

In line with the permaculture principle of beneficial relationships, we can carry out a systematic input–output analysis of the animal to ensure that we make the best use of all the animal's functions.

The chicken is originally a jungle bird that lives in small flocks of 25–30 animals, of which approximately 3–6 are roosters. If you keep more chickens together than there are in a natural

The five-year-old self-feeding chicken forest includes Siberian pea trees, sea buckthorns, Oregon grapes and blackcurrants for chicken feed, as well as hazel and pear. Jerusalem artichokes are one of the robust plants that can grow in the ground cover. Reforest Farm, Denmark.

flock, the chickens will not be able to keep track of each other and establish a natural pecking order. The result will be stressed chickens and everyone will peck at everyone.

Hens and roosters clean their feathers by dust bathing and need a place where they can do so. An old rooster walks in the middle of the flock and finds places with food, nesting sites and sleeping places and mates with the hens. The young roosters go to the edge of the flock and defend the flock. If we take the roosters out of the flock, the hens will not forage as far to find food, as that is the rooster's job.

The hens scrape for seeds, insects, worms and plant sprouts for food, lay eggs, incubate them and look after the chicks. The chicks stay with the mother, sleep under her wings and eat from her beak.

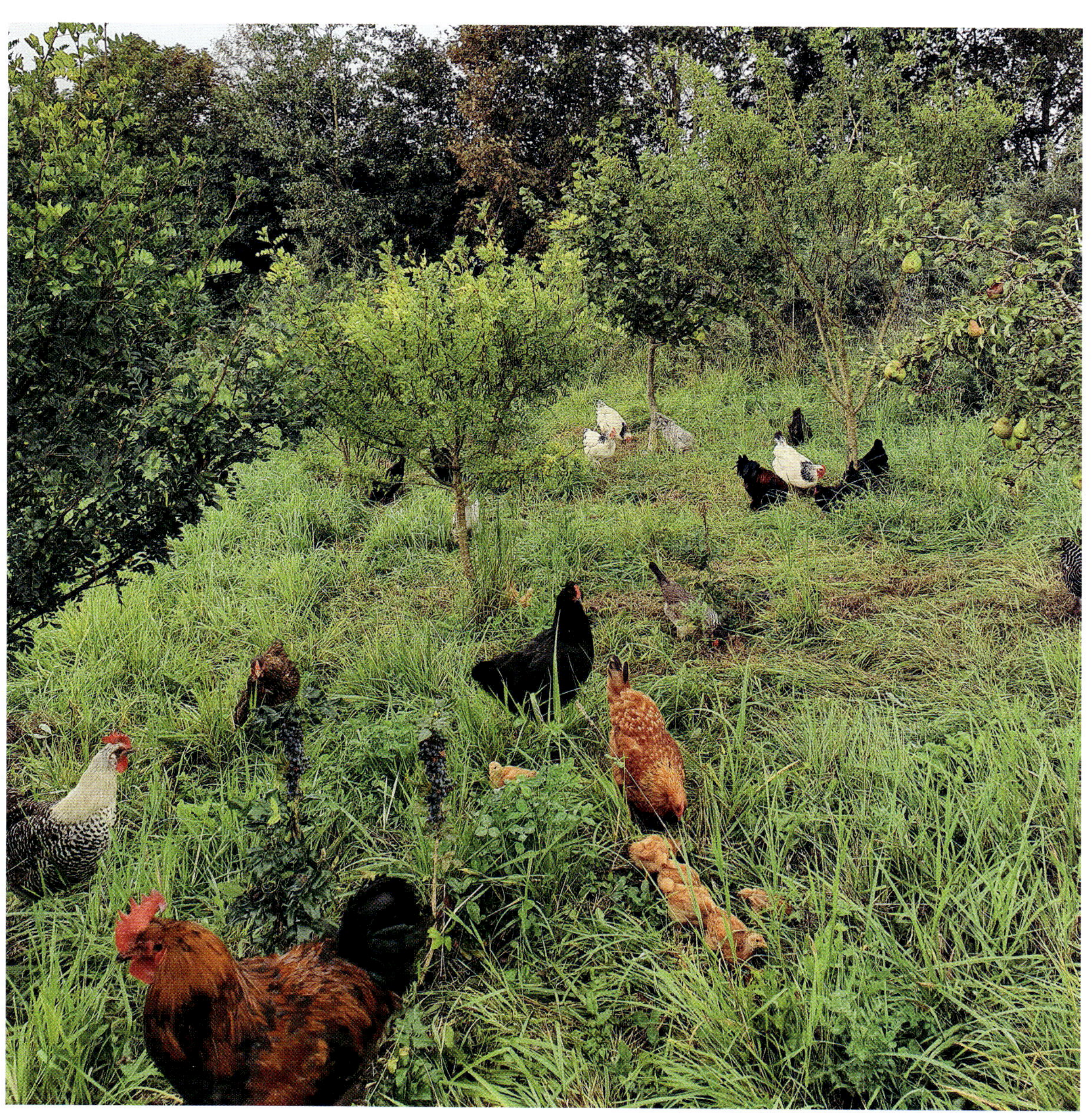

Chickens are intelligent animals with family relationships, empathy, a sense of time, logical-mathematical intelligence, many different sounds for communication, etc.[18] The chickens learn from their mother and therefore become better at finding food in nature.[19]

When they are about 20–24 weeks old, the chickens begin to lay eggs. The intensive breeds are bred to lay large eggs and up to one egg per day, and to start laying eggs at a younger age.

The large, frequent eggs and early egg laying cause the chickens to break their keel bone again and again, and this is painful.[20] This is new knowledge, and research into the problem is not complete.

Many of the intensive breeds have also lost their instinct to go broody and thus be able to incubate eggs and be mothers. If we want a chicken farm where the flock is constantly renewed, so that we are not dependent on buying new chickens, we must therefore choose a heritage breed. Based on this knowledge, we can do an input–output analysis:

Inputs/needs:
Roaming under trees
Sleeping high
Dust baths
Scratching
Nesting
Rooster
Food, with protein and fat and fresh shoots

Outputs/functions:
Eggs
Meat
Dung
Scratching
Feathers/down
Chickens
Heat
CO_2 from exhalation

Chickens

How can we design based on the primeval forest and the chickens' input and output? Below are examples of permaculture systems with forest and chickens.

The self-feeding chicken forest

We can design a self-feeding chicken forest, where the chickens are allowed to roam in a protected environment under trees that have been specially selected to produce food for them. Fruit trees can be planted, which are primarily harvested by humans, but where the chickens will take what falls. The following is written based on our own experiences, with new ones being added all the time.

It is an advantage to divide the chicken forest into several pens, which can be used in rotation so that the 'forest floor' stays green and insects and worms can reproduce. The trees are planted at a distance, as in the food forest, that allows light to reach the bottom.

Among trees and shrubs that yield food for the chickens, the nitrogen-fixer Siberian pea tree, which produces protein-rich seeds, should be mentioned. And the nitrogen-fixing sea buckthorn, whose berries are rich in oil. Strongly colored berries such as elderberry, blackcurrant or chokeberry contain many vitamins and antioxidants that strengthen the health of the chickens.

Garlic and Jerusalem artichokes can be planted, which will be able to shoot several times from their bulb and root tubers, even if the chickens scrape and eat the shoots. Chickens also like to eat seeds and leaves from the perennial vegetable Turkish rocket.

When chickens eat insects and worms, they utilize an otherwise unused output. In some cases, these are pests such as codling moth and hazelnut weevil that cause worms in apples and hazel. In this way chickens have an additional function in keeping the cultivation system healthy.

We recommend that a special winter forest be created, where evergreen shrubs are planted to increase the well-being of the chickens, which provide shelter in winter. Here we suggest the shrub Oregon grape (*Mahonia aquifolium* where the dark berries still remain on the bush in winter for winter food, and the nitrogen-fixing Ebbinge's silverberry, where they eat the leaves as winter greens.

The ground in such a fold will inevitably become bare over the winter, and then it can be sown in the spring when the chickens are moved to a summer chicken forest. The chickens have made a seedbed with their scratching, which can be sown with grass, grain or sunflowers, that is raked over. The plants will be ready with feed when the chickens are put back to this fold in the fall.

In the tropics, self-feeding chicken forests can produce all the chicken feed all year round, but in temperate climates, we must supplement with cultivated and stored winter feed. Kitchen waste and propagated worms from a worm compost are also good feed supplements.

A self-feeding chicken forest can be large. We suggest around 40m^2 per chicken, but there is a lack of experience in the area. It may sound like a lot, but it will reduce the area that is otherwise used for feed production somewhere other than the chicken coop.

The chicken tractor

A chicken tractor is a movable enclosure with or without a housing unit. On a garden scale it is built for a few chickens. With the chicken tractor we can let the chickens harvest their own feed, utilize their scratching behavior and their manure. It can be moved to where we want a bed weeded and cleared of weed seeds, pests and disease before we sow a crop. It can be part of a raised-bed garden or a no-dig garden if it is built so that the bottom fits the beds. The chicken tractor has some limitations. The chickens will compact the soil a little, as they only have a small area in relation to their needs, and it is time-consuming to move the chicken tractor. Therefore, they should also have a permanent chicken run.

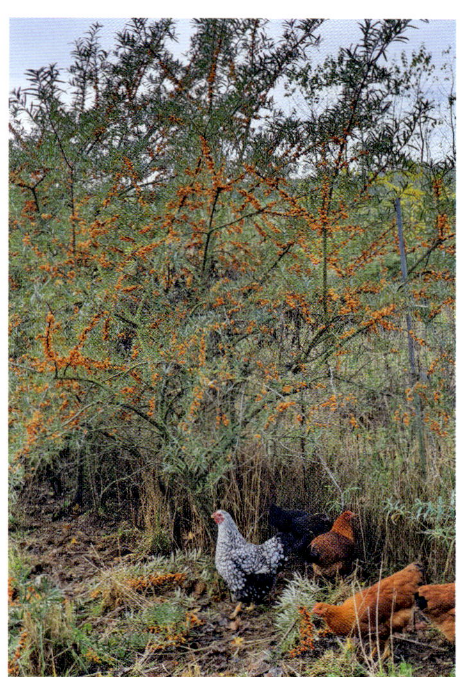

The berries of the sea buckthorn are rich in vitamins, oil and protein that chickens need to be able to lay eggs. Late in the fall the berries are ripe and the branches can be cut off and used as feed. The chickens harvest the lower berries on the shrubs themselves. Reforest Farm, Denmark.

Chickens in crop rotation

Chickens can be part of a crop rotation with vegetables so that each year they go to a bed or a field before a nitrogen-demanding crop, such as cabbage or potatoes. The chicken coop can be advantageously placed in the middle of the cultivation area with several paddocks extending from it, for example, one for chickens, one for potatoes, one for cabbage and one for grains or beans.

The chicken greenhouse

Here, the greenhouse is built together with the chicken coop, so that the greenhouse faces south and the plants get sunlight for photosynthesis. The chicken coop is built on the north side.

The wall between them must be built from a heat-conducting material, so that the heat from the chickens at night will give a higher temperature in the greenhouse. This, together with the chickens' exhalation of CO_2, will increase the growth of the plants in the greenhouse. In the heat-conducting wall there must be an opening at the top for the exchange of CO_2 and heat, but with a dense fabric lining to prevent dust from the chickens' dust bathing from getting onto the plants. Openings can be made at the bottom of the wall so that the chickens can come in and weed and fertilize the greenhouse when desired.

The movable chicken coop

On a permaculture farm, we can also make one or more movable chicken coops that can be moved to the kitchen garden, the meadow, the forest and the self-feeding chicken forest, where a temporary fence is erected. In this way the chickens will have a much larger area to use and get feed from. This is really just a bigger version of a chicken tractor. The chicken coop is made with wheels, so that it can either be moved with a small tractor, a draft animal or, if it is a very small house, on a wheelbarrow. If the house is moved daily, the bottom can be made out of wire mesh, so that the manure falls directly to the ground and we avoid the work of removing it. Instead we fertilize the different places where it is placed. Chickens poop most at night, and it is a trade-off whether we prefer to be able to collect the manure to use it for the intensive cultivation areas. The disadvantage of this system is that moving chickens and pens is time-consuming and can stress the chickens.

The large wheels make it easy to move the chicken coop, so the chickens can be moved around between the rows of trees in the planned alley cropping at Calendula Permakultur, Denmark.

Signe Gerlach keeps geese that graze under six-year-old cider apple trees at Underwood, Denmark.

Ducks and geese

Ducks and geese have many of the same functions as chickens, as they produce eggs, meat, down, etc. Ducks are often included in food forests, forest gardens or vegetable gardens during the growing season: they do not scrape like chickens, so they can be let into a cultivated and mulched bed and find slugs that would otherwise destroy our crops. Only the newly established annual vegetables must be fenced off from the ducks. Ducks can also be kept as slug fighters in more simplified systems just with fruit and berries.

Slugs can be a particularly big problem in permaculture systems, where, in contrast to a plowed winter field, there are many hiding places all year round.

Indian Runner and Khaki Campbell ducks are two breeds that are especially good at regulating the slug population because they prefer slugs to vegetables. Both breeds are prolific egg layers, and they generally don't fly although the Khaki Campbell sometimes get into the habit of fluttering over low fences.

Muscovy ducks are larger birds, making them better for meat production, but they don't lay as many eggs as the Runners and Campbells. Muscovy ducks

Muscovy ducks with ducklings in the no-dig market garden on the Calendula Permakultur Farm, Denmark.

REGIN BENNETZEN

ESBEN SCHULTZ

The Indian Runner duck is a small bird that was originally bred to catch slugs in rice-farming systems and is therefore the most effective for slug control. They can roam in vegetable gardens, agroforestry, orchards and forest gardens. It is advantageous if they are completely tame; they can be moved where needed and to safety from the fox at night.

The ducklings are fed only by hand for the first two weeks and a chosen 'calling sound' is made at the same time, such as 'come, come, come'. When you make this sound later on, they will come to you for the rest of their lives. Reforest Farm, Denmark.

are more prone to go broody and have their maternal instincts intact, which is not the case for the Campbells and Runners. A Muscovy duck will be able to hatch and raise Campbell and Runner ducklings just fine.

Ducks do not need much food beyond what they find themselves, but a little seed to keep them tame and get them into their duck pen and house is useful. Ducks need water to clean their beaks and swim in.

With ducks we have an excellent example of how we can work with nature and create a *beneficial relationship*. With good design, we can turn the slug problem into a food source for the ducks, which then provides an extra benefit from the system. Instead of collecting slugs, making slug barriers or using poison, we create a self-reliant and highly productive system.

Geese eat grass and tender shoots. They can be used as weeders because they avoid many broad leaf plants. They have been used as weeders in fields with strawberry, potatoes, beans, onions and sugar beets.[21] We can train them to be good weeders when they are young if we only present them to grass and weeds, and not to our crops. Geese can also be kept in fruit and nut forests or on lawns where fresh grass can be their main source of food. If we plan to keep geese in an orchard, we must be aware that they will sometimes strip the bark of young trees, so the trees must be individually protected e.g. with chicken wire.

Pigs

Pigs are omnivorous, just like humans, which means that they compete with humans for food on a global scale. Even in a local permaculture system, they will often eat food that could have been eaten by humans.

JIM TERP

The pigs at the farm Den Gode Gård live primarily in the forest, which is the pigs' natural habitat. They dig deep for roots, earthworms, larvae and other goodies, and it is clear that they enjoy it. They also have pasture available, because it is healthy for the animals to eat fresh grass. The pigs are moved at regular intervals to get the right balance between healthy disturbance of the soil and destruction. It is rejuvenating for the forest to have some disturbance, but it must be in moderation so that it benefits the soil, the forest and the pigs. Denmark.

The pig is a forest animal, where the females live in small herds of two to eight. They rummage in the soil for food with their snouts. This 'snout power' can be used for soil preparation and weed control. Depending on the breed, they will work the soil down to a depth of 70cm and remove root weeds such as couch grass. That is why they are used in permaculture systems for 'pig plowing' before annual crops, which are grown on slightly larger areas, such as grain, cabbage and potatoes.

The pigs must then live in small pens that are moved frequently to ensure that the soil preparation and fertilizer are evenly distributed over the area. It should be noted that pigs will cause soil compaction if they are put out on wet soil.

This compaction can be used constructively when you want to build a pond. When their snouts rub back and forth and since there is a lot of weight on small pig feet, they will compact clay soil and thus help to make a pond watertight.

The pigs must have a permanent pig pen somewhere else, otherwise they will compact the soil where we do not want it. The permanent pen should be planted with trees so that the pigs will have shade, and the nutrients are taken up by the trees. This is an example of the principle of stacking. Energy crops such as poplar and willow will grow well from the pigs' manure. The trees can be kept as pollards, where the branches start at a height where the pigs cannot damage them.

Under fruit and nut trees, pigs, in addition to fertilizing, will be able to eat fallen fruit and nuts, which, together with the animals and roots they find in the soil, will provide a partial self-feeding system. Another useful function here will be that the pigs keep the population of voles down.

Pigs can only be introduced into such forest systems when the trees are of a certain size and can withstand the pigs rubbing themselves on the trunks and rooting around the roots with their snouts.

In order to go as far as possible in creating a system that is not dependent on external input, we can choose a traditional breed for permaculture systems. Traditional breeds are better foragers and need less bought-in feed. The breed Kunekune can even be raised and finished on good-quality pasture without the need for supplementary feeding.[22]

The modern pig breeds are bred to grow quickly; they will be constantly hungry and will not thrive on a natural diet, while more traditional pig breeds will simply grow more slowly. Modern pigs have been bred to have three litters per year, whereas traditional breeds only had one litter per year. This allows for meat production that follows the course of the year.

Ruminants

Common livestock such as cattle, sheep and goats are all ruminants. Horses are not in terms of digestion, but they eat the same food and can be part of the same farming systems. Ruminants can break down highly fibrous feed, such as grass, straw, leaves and bark – plant material that is otherwise not utilized. This means that they can often be introduced into our permaculture farming systems and provide additional by-production that helps increase the overall yield.

Climate issues

It is an inconvenient truth that ruminants contribute to climate change because they burp and fart methane when breaking down the fibrous feed. Ruminants are part of our culture, and many of us value the products in the form of meat and milk. Ruminants are also important on most organic farms because the grass-clover ley that is part of the crop rotation helps maintain soil fertility without the use of artificial fertilizers.

There is nothing wrong with ruminants in themselves; there are also ruminants in nature. The problem is that there are too many of them today. There are now about six times as many large ruminants on Earth as there were just 500 years ago, when you include both wild and domestic animals.[23] Globally, ruminants help to throw the climate out of balance.

Methane stays in the atmosphere for a shorter time than CO_2 but is much more powerful in terms of the greenhouse effect, especially in the first years after it is emitted. Methane is 28 times as powerful as CO_2 measured over a 100-year period, but over a 20-year period it is a full 80 times as powerful. The effect of methane in a 20-year perspective is essential to consider, as climate research indicates that we must reach net zero emissions by 2050 to ensure a stable climate. Reducing the number of ruminants can be seen as pulling on a climate emergency brake.

It is possible to store carbon in permanent grasslands, but this hasn't been fully researched, so there is discussion about how great the potential is.[24] However, it is estimated that grasslands store a maximum of carbon equivalent to 3 tonnes of CO_2e per hectare per year, and usually significantly less.[25, 26]

One cow produces approximately 100kg of methane per year,[27] which over a 100-year period has a climate impact equivalent to 2.8 tonnes of CO_2e per year. Measured over a 20-year period, the 100kg of methane corresponds to 8 tonnes of CO_2e per year.

Methane emissions per hectare of course depend on how many animals we have per hectare. As an example, we can imagine that we have two cows per ha of permanent pasture. The climate impact from the cows will be 5.5-16 tonnes of CO_2e per ha per year depending on which conversion factor is used for methane.

Overall, the negative climate impact from a permanent pasture with two cows per hectare is therefore between two and five times as high as the positive, and this

KRISTINE DALBACH

With holistic grazing, the cows are confined to a small paddock, so they walk closely and graze evenly, like a herd wandering across a plain. Since weeds are not multiplied and the grass has good regrowth, the field can be kept as permanent pasture so that carbon is stored in the soil. Jersey cows at Søgård Cooperative Farm, Denmark.

is when we achieve maximum carbon storage in the pasture.

On top of this, there are actually methane and nitrous oxide emissions from the cow's manure, as well as the climate impact from the production of feed, barn buildings and processing of the products, etc.

Even though we have cows on permanent pastures, the climate impact from cattle will usually far exceed the climate benefit from carbon storage in the grassland. From a climate perspective, we should therefore prefer to create our carbon-storing agricultural systems without ruminants.

Although ruminants must be limited in our agriculture, we must also realize that they have an important role to play in preserving and restoring biodiversity, where grazing helps maintain open and semi-open habitats.

In some cases, horses can replace ruminants when we need grazing for biodiversity purposes, and here the climate impact is somewhat lower. An average horse only emits approximately 21kg of methane per year.[28]

Below, we will review different ways of having ruminants that follow permaculture principles, but it is also crucial that there is a sharp reduction in the number of ruminants here and now. Therefore, we also propose that they be greatly scaled down and primarily used for biodiversity purposes in our proposal for land use, later in the chapter (p.124).

Co-grazing

By allowing ruminants of different species to graze together, we can achieve a number of *beneficial relationships*. They graze together either by being in the same paddock at the same time or by different animals grazing after each other. Dairy cattle, which need a lot of protein in their feed to be able to produce their daily milk, enter the pasture first. Cattle will not graze around their own cow pats, but when we let another ruminant or horses come after the cattle, they will graze close to the cow pats because they do not share parasites with cattle. Sheep and goats share the same parasites and should not graze in the same system.

Sheep graze low, cattle and horses graze at medium height, and goats are top grazers, nibbling only the top of the sward. Combining these different ways of grazing will provide better utilization of the grass.

Holistic planned grazing

Holistic planned grazing is part of a wider framework called holistic management. It is about carefully planned movements of large herds of livestock in a way that mimics the processes of nature. In a context of organic farming where pasture fields are normally included in the crop rotation, the scope of holistic planned grazing is that it allows us to keep permanent pastures and maintain high yields over the years, while at the same time we store carbon in the soil.

In nature, ruminants migrate in dense herds to protect themselves from predators. The animals graze all plants in an area uniformly and thus also eat less attractive plants, such as thistles, but do not manage to bite the most attractive plants all the way down to the ground. The animals move on in flight from predators, and the area is left for a period of time.

When biting off the grass, part of the grass roots will die and leave dead plant matter in the soil, i.e. leaving carbon behind. The part of the grass root that is still alive is about as large as the part of the plant that remains above the ground, and since it is a relatively tall tuft of grass, the root will be larger than if the grass

In nature, birds follow the ruminants and find parasites in their manure; this can be imitated in holistic grazing. The mobile chicken coop is moved into the field 3–4 days after it has been left by the cows. Søtoftegård, Denmark.

MIKKEL POULSEN

had been bitten off all the way down. The grass regrows well because the herd of animals has left a lot of manure and because the plants have a large root to grow from.

This is in contrast to the way ruminants have been kept in agriculture for centuries. The animals are kept in the same large paddock for a long period. Since there are no predators in the paddock, they will spread out and graze selectively, whereby the most attractive plants are completely grazed down and have little root left, and weeds are left behind and multiply. Therefore, the farmer must plow the pasture and resow it, and the pasture is included in a crop rotation. As mentioned earlier, plowing results in over-aeration and loss of carbon.

When we imitate nature and instead create small paddocks where the fence is moved every day, we can take the place of the predator and achieve more uniform grazing. The system must be managed so that the ruminants do not graze too far down so that the grass roots remain strong.

The method was developed by Allan Savory for cultivating drylands in Southern Africa. Here, the large amount of fertilizer that falls densely, which contains seeds and water, will fertilize the soil. Thus, there are examples of infertile dry areas that have today become grasslands.

In recent years, the method has gained popularity all over the world, in completely different climates, such as in Scandinavia and Great Britain. The first experiences with holistic planned grazing with dairy cows in Denmark indicate that one must accept a lower milk yield, since the feed is not as optimized, but in return, the operation can be done with lower input and costs.

In nature, birds will follow the ruminants. The birds will distribute fertilizer evenly in the area with their scratching, and if the ruminants have parasites that come out with the fertilizer, the birds will eat them and thus keep the ruminants more disease-free. Therefore, there are several examples of farmers who incorporate the mobile chicken coop into the holistic grazing system and let the chickens run there for such a short time that they do not have time to scratch the soil bare.

In this system, egg-laying hens will need protein and oil-containing feed from another place and will thus add even more fertilizer to the grass, and we must be careful not to overfertilize.

Foggage

The foggage system is a traditional way of feeding beef cattle and sheep, which has been further developed by the Hollins family at Fordhall Farm in Shropshire, England.

Foggage is grass that has been left from late summer to be eaten in winter. The beef cattle and sheep graze outside all year round, and this saves work and energy consumption for buildings, production of winter feed and hauling out of manure. The grass must have a well-developed root system to prevent soil compaction.

The limitations of the system are that it only works on very light sandy loams, otherwise the soil will easily be compacted in wet weather during winter, and that the grass in winter does not contain enough nutrients to be used as feed for dairy animals, and animals for meat production will grow more slowly.

Natural grazing

With natural grazing, the primary purpose of keeping animals is to support biodiversity so that we achieve a high species richness of different plants in the grass and of the insects that are associated with the plants and the animals' dung. Over time, natural grazing will lead to a rewilding of the landscape.[29, 30] The production of food is secondary here.

Some central principles of natural grazing are that the animals are kept outdoors all year round, as in the above-mentioned foggage system. The number of animals per unit area is set so that it corresponds to the density in a natural ecosystem, that is, it is adapted to the amount of food the animals can find in winter. Ideally, the animals are not given supplementary feed, but we can of course do so in particularly harsh winters.

With this population density, the grass will be completely eaten down at the end of winter, and therefore it will be possible for seeds to germinate and take hold in the spring. Conversely, the animals cannot keep up with eating the grass in the summer, which means that the plants have ample opportunity to flower and set seed. Today, the number of grazing animals in the wild is generally lower because of intensive hunting.

With natural grazing, animals should preferably have large areas to move around, and over time the succession will lead to a mosaic landscape, like the one we described in the section on the primeval forest as a model.

Polycultures of grass and trees

In parts of the world where natural succession leads to forest, the above grazing systems can advantageously be combined with trees. Grasslands in which trees are planted store roughly three times as much carbon as grasslands without trees, but of course it depends on how many trees are included in the system. They will therefore be more effective in counteracting climate change, while at the same time being better able to withstand drought due to a deeper root system and because the trees reduce drying out from the wind. The roots of the trees and the associated microorganisms are active even in winter and can therefore retain water and nutrients.

Polycultures of grass and trees are a good example of stacking. Grass growth peaks at the start of May and will have had a lot of its growth before the trees come into leaf and take the light. Such systems make it possible to mix in nitrogen-fixing trees and will provide a good development of mycorrhizal fungi,

JOEL HOLMDAHL

Sheep graze on silvopastures, where the trunks are protected by prickly juniper shrubs. Rikkenstorp, Sweden.

which can be associated with both trees and grass. This can make the nutrient cycle more stable (see the section on nutrients, p.111).

These polycultures can be divided into two types:

Silvopastures, where the animals eat leaves from, for example, ash, lime, poplar, holly, alder and mulberry trees. The trees are established either along the edge of the field, from where we can throw the branches into the paddock when pruning, or the trees can be planted as individual trees in the field and pruned to a height that means the animals can only reach the lowest branches. Silvopastures will increase the total amount of feed, increase the well-being of the animals by providing shelter and shade, and produce wood.

Orchards, where animals are kept under production trees such as tall trees for fruit, juice or nuts. Here the main harvests will come from the production trees, and the animals should be thought of as an additional yield, a by-product that

SANDRA VILLUMSEN

Shropshire sheep are frugal and do not bite the bark. They can therefore graze under the cider apple trees, and thus the area is utilized in several layers. Farendløse Mosteri, Denmark.

In the newly established nut forest, the function of the goats is to maintain the grass and provide a by-product of meat. The walnut trees are protected with a small 'greenhouse' that promotes growth in the first few years. Reforest Farm, Denmark.

also provides benefit by fertilizing and keeping the grass down.

Alley cropping

Both silvopastures and orchards can be systematized by planting the trees in rows so that there is space for machinery to drive between the trees. In countries such as France, where there is sufficient energy in the sunlight, there are examples of annual crops, such as grains and vegetables, being grown between rows of production trees.

Further north, however, light is a limiting factor, and annual crops will grow more slowly. Grain might not get enough light to ripen its seeds, and it can also be more uncertain whether the grain will be dry at harvest time when it is grown in the shade under trees. Perennial vegetables or grasses are better adapted to this niche. While the trees are small, there will be enough light to grow annual crops, so here it is important to remember the permaculture principle of *succession.*

Alley cropping eliminates the labor-intensive protecting of trees and allows for machinery access. If we do not want fencing or machinery in the system, there is no real reason to plant the trees in rows.

On land with a steep slope, the trees are planted in rows that follow the contour lines. This prevents erosion, as the tree roots and leaf cover help to keep the soil permeable, allowing water to seep down. On land with a limited slope, the trees are planted in a south–north direction to get as much and as uniform light as possible for both the trees' ripening of fruit or nuts and for the growth of the plants under the trees. One side of the tree crown will then receive the morning light and the other side the afternoon light.

Examples of trees for agroforestry

Trees such as chestnut and walnut can replace some of what we currently think of as essential staple foods, i.e. potatoes, cereals and rice, so we will need less field crops and instead get a larger part of our diet from carbon-storing forests. They can be planted in polycultures with tall nitrogen-fixing trees, and pigs, geese or chickens can roam under them.

Chestnut, full of starch and omega 3

If you use the right varieties, it is possible to grow chestnuts in temperate climates where summers are short. Chestnuts have a yield of 2.8–5 tons/ha. Of the varieties, Bouche de Betizac is particularly recommended in temperate climates, as it has a high yield every year.[31] It is recommended that it be planted with other varieties, as they cross-pollinate; for example, the varieties Marigoule or Marron de Lyon. They cast a dense shadow and are therefore difficult to underplant. Nutritionally, they are similar to potatoes in protein but contain twice as much carbohydrate and eight times as much fat, which includes healthy omega 3 fatty acids. They contain vitamins A, B, C and E, a wide range of minerals and fiber.

Walnut, high yield, rich in protein and omega 3

Modern varieties begin to bear large, tasty nuts which are easy to crack, after six or seven years. There is a wide range of recommended varieties, and it is best to have several varieties for pollination. Walnuts provide dense shade and have a mild growth-inhibiting effect (allelopathic) on many other species. Therefore, they are difficult to underplant in a polyculture. The yield from mature plantations is 3.5–5.2 tons/ha/year.[32] This is similar to the

The 13-year-old chestnut tree of the Marigoule variety is now bearing nuts abundantly at Reforest Farm.

9-year-old walnut tree of the Geisenheim variety in the nut forest at Reforest Farm.

yield from chestnuts, but the content in walnuts is more concentrated, with a high level of omega 3 fatty acids.

Paulownia – the world's fastest growing tree

Paulownia tomentosa originates from Asia, and hybrid varieties have now been developed that can withstand temperatures down to -40°C, so it can be used everywhere in temperate climates.

After just 10–12 years, the trunk reaches a diameter of 35–45cm, and the tree can be felled. Although the wood is light, it is relatively strong and hard and can be used for construction. After felling, the tree shoots again from the stump, and it can do so up to seven times.

Because the tree is so fast-growing, it is also called the 'climate tree', as it is believed to be able to draw large amounts of CO_2 out of the atmosphere. A calculation has shown a storage of as much as 52 tons CO_2e/ha/year.[33]

Because paulownia grows so rapidly, it may be useful as a nurse tree that can quickly establish a forest environment in a new agroforestry system. However, we must remember that it casts a lot of shade. The leaves, which with a width of about 50cm are the largest leaves on any tree we can grow in a temperate climate, are rich in protein, fat, sugar and the nutrients nitrogen, phosphorus and potassium. An 8–10-year-old tree sheds 100kg of leaves in the fall, which can be used for fertilizer, feed and biogas production.

Paulownia is new to northern Europe and North America, and we do not know how it grows in these parts of the world.

The wood of the tree is somewhat brittle, which means that branches fall off in strong winds. Will it be windproof enough where climate change will lead to stronger winds? We also don't know if it will have problems with diseases.

When the species is used in areas where it is not native, it is also very important to use only sterile clones that do not spread by seed. If it spreads in the wild, it will, as a fast-growing pioneer plant, quickly outcompete the native plants, which will be very destructive.

It is an exciting species to try in our permaculture systems if sterile clones are used, especially because of the great potential for carbon storage.

Rabbits

Rabbits are pseudo-ruminants and therefore eat the same food as ruminants. In a garden or a small farm they can digest the fibrous feed, and at the same time we will have meat production based on a self-feeding system. Rabbits breed and grow quickly and will therefore provide a lot of delicious meat.

Rabbits can, for example, be kept in a 'rabbit-lawn mowing cage', which is built to maintain lawns or grass paths. The cage must have a certain size to allow room for movement – depending on the breed, rabbits can jump up to 8 meters. A good design includes wheels at one end, so it can be easily moved to fresh grass every day. Since rabbits dig, there must be metal mesh at the bottom. This is best done by laying out the netting where the cage is to be moved, because then the grass will grow through it, rather than being laid flat when the cage is moved.

Rabbits can also be kept more naturally if a fence is dug along the edge of a larger paddock. Here they will dig tunnels and make burrows where they live in families.

Honeybees

With beekeeping we can produce honey and ensure pollination of fruit trees and other crops. Because honeybees compete with wild bees for food, we must be aware that beekeeping on an excessive scale can impact wild bee populations, many of which are endangered.

By creating cultivation systems with high diversity, we increase the food base for both wild bees and honeybees. The best pollination is achieved by having different species of bees present.

Beneficial wildlife

Some wild animals are beneficial on our farms and in the garden, where they help create a balanced ecosystem, and we can attract them in various ways. Slugs and snails are predated by toads, frogs and hedgehogs. We can actively support them by, for example, building ponds to provide habitat for frogs and toads, and making brush piles to provide habitat for hedgehogs.

JIM TERP

At the farm Den Gode Gård, the rabbits live close to nature and dig their own tunnels where they breed. There are about 50 females, and the two pens, between which they change every year, are each 1 hectare. In addition to their grazing, the rabbits are fed in a pen with branches, kitchen garden waste and grain as well as hay in winter. Denmark.

We can plant bee plants that provide food for wild and domesticated bees throughout the season, so we get better pollination. We can also make 'bee hotels' for wild bees and although it is a permaculture principle to keep the soil covered, it is also a good idea to make sure there are some small areas of bare loose soil, preferably sunny places, where ground bees can build nests.

Stone piles in sunny places are habitat for stoats and snakes that eat mice and voles, as well as habitat for lizards and some insects.

We can also put up bird boxes and attract birds that help control insects and provide an influx of fertilizer that can be considerable.

ANNEVI SJÖBERG

Beneficial insects can be supported by making insect hotels from different materials and with different holes and depths. The hedgerow is also a habitat for beetles, hedgehogs, etc. Permaculture Stjärnsund, Sweden.

By building ponds in gardens and on farms, we create habitats for a wide range of beneficial animals, such as frogs, toads, newts, insects, etc. The picture shows the pond in the food forest at Reforest Farm, Denmark.

Birdhouses can be set up to attract small birds. Krameterhof, Austria.

Grain cultivation

As was evident from the review of the primeval forest as a model for our permaculture systems, annual plants only have a small place in the natural ecosystem. Because many of these plants have been bred over thousands of years – to enhance flavor, pest resistance, durability and ease of growing – they are worth including in permaculture systems. This refinement has not occurred in many perennial vegetables or nut trees.

Cereals are annual plants, and they are grown in monocultures on huge areas that are plowed and fed with fertilizers and pesticides. This production creates high greenhouse gas emissions, leaching of nutrients into the aquatic environment and loss of biodiversity.

Grain was the first crop that humans began to cultivate systematically approximately 12,000 years ago in the Middle East. All the cereals that are grown today originally come from wild grasses in the Middle East and are adapted to this climate. This means that, the winter, when the grain grows, is wet, and the summer, when the grain ripens and withers, is dry. When cereals are instead grown in areas with rainfall throughout the year, such as northern Europe, there will be difficulties in harvesting the grain when it is dry, and we will have the problem of nutrients leaching into the aquatic environment.[34]

Culturally, however, many of us will find it difficult to imagine our diet without grains, and therefore we have to deal with how we replace grains with perennial crops or how we grow them in a responsible way.

Less consumption of grain

The most straightforward and most effective way to avoid the high costs of grain cultivation for the environment and climate must be to use less grain.

When we look at agriculture today, the cultivation of cereals and other crops, such as rice, maize, potatoes and millet, takes up 45% of the world's cropland.[35] These are crops that all have a high content of carbohydrates and a relatively low content of oil and protein, which are characteristics that are not found in many perennial crops in temperate climates.

A large part of the grain that is grown is used as animal feed. By refraining from eating meat and animal products altogether, eating them minimally or eating those that are produced on permaculture farms where grass and other perennial plants are used as feed instead of grain, we are helping to limit grain cultivation.

If we look at Denmark, as an example, grain is grown on 61% of the plowed agricultural area, and by far most of the grain is used to feed the many animals that are produced for export.[36]

A normal consumption of bread and other cereal products is around 200g per day, and this can be covered by cultivation on only 146m^2 of agricultural land with organic cultivation methods. The grain Danes eat can be grown on 3.5% of the plowed area.[37]

Part of our consumption of cereal products can, as previously described, be replaced by chestnuts, which have historically functioned as a staple food, e.g. in some areas of the Mediterranean. Chestnuts can be ground into flour; however, it does not contain gluten and must be mixed with ordinary flour if it is used for bread that needs to rise.

Grain cultivation and efficiency

The large spread of grain cultivation is, among other things, a result of the fact that it is relatively easy to mechanize. In the industrialized part of the world, only about three working hours are spent on a farm to produce 1 ton of wheat.[38] Added to this is the labor time spent outside the farm for the production and maintenance of machinery, the manufacture of chemical fertilizers and the extensive distribution system that is necessary when doing such large-scale production.

At first glance, the industrial cultivation of grain is an economic and labor-efficient system, but this is only when we do not take the costs to the environment and the climate into account. When we are attempting to increase biodiversity and reverse climate change, it is necessary that the huge areas that are now used for grain cultivation be converted to carbon-storing cultivation with high diversity.

Returning to old farming methods with manual labor and animals as traction is not a good solution, as they were also based on plowing. In addition, there is a large amount of labor associated with such a form of cultivation. In the year 1800, it took about 170 hours of hard work to produce 1 tonne of grain,[39] i.e. about 57 times as long as using industrial agriculture methods. Permaculture must find a third way, without the environmental burden of industrial agriculture and without a return to the hard work of the old days.

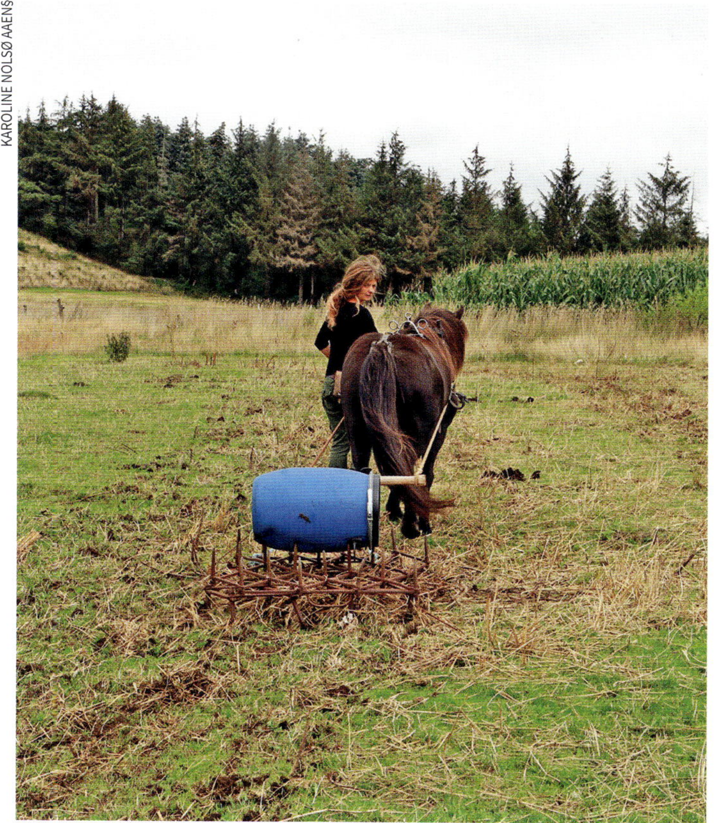

The grain is sown and then harrowed into the permanent pasture at Myrrhis Agroforestry, Denmark.

KAROLINE NOLSØ AAENS

Grain cultivation in permaculture farming

If we reduce the consumption of grain drastically, the production that remains will be able to fit into the landscape based on a permaculture design, unlike today, where in many places in the world it completely dominates the landscape.

On the one hand, in terms of resilience, it is a good idea that we maintain knowledge and skills about how we can locally produce the grain we need without the use of machines or other inputs. But in order to reduce the amount of work, it would be best if most of the grain production were done with the use of machines.

Overleaf are examples of cultivation methods that follow permaculture principles as much as possible, but that can be used on both a large and small scale.

Some of the examples are challenged by significantly lower yields, and this conflicts with the principle of *small intensive scale*.

On Svanholm, the ley is terminated with a shallow cultivation, and at the same time preparations with effective microorganisms are sprayed on the plant matter to start the composting process. The grain is then sown. Denmark.

Grain in a living cover crop[40,41]

For several decades, organic cultivation methods for cereals in a permanent cover crop in the form of white clover or micro clover have been developed. The grain is sown either with a special seed drill with discs that cut slits in the clover to make it possible to sow directly into it or with a seed drill with rotary cultivators that clears strips in the clover in which to sow. In some cases, the clover is cut back by grazing it with sheep immediately before sowing. This also has the advantage that the sheep's manure quickly fertilizes the soil for the grain. In other cases, the clover is cut back mechanically by mowing.

Since clover fixes nitrogen from the air, the system can largely supply itself with nitrogen. The absence of plowing and the presence of a permanent root system also means that there is no risk of nitrate leaching. Less diesel is used, especially because there is no plowing. When the soil is not disturbed, the soil structure can be improved year after year, and the content of organic matter in the soil will increase, which in addition to increased fertility also means that carbon is stored so that climate change is negated.

This method has been tested in both organic and conventional experiments. The yield of grain in successful organic trials has been 70% of the yield in control plots with plowing. In addition, there is a possible yield of clover for the sheep. The lower yield is due to the clover competing with the grain.

A problem with growing cereals in a permanent ground cover of clover is that over the years the clover will be infected by perennial root weeds. It's a problem that will be difficult to solve without plowing the field every now and then.

In Denmark, Tycho Holcomb and Karoline Nolsø Aaen have developed a variant of cereal cultivation in a living cover crop, where the ground cover consists of a permanent pasture.[42] The grass is set back by letting horses or chickens stay on the field, before sowing at the end of August. After sowing, a light horse-drawn harrow is pulled across the area to cover the seeds. The harvest is done with hand tools, and they grow only a small area for their own consumption of rye for bread. The yield corresponds to around 1.5 tonnes per hectare, i.e. approximately 30% of the yield in regular organic grain cultivation. Added to this, is the yield from grass. Since the method can work even if there are perennial weeds in the grassland, grain cultivation can in principle continue indefinitely without plowing.

Grain in mulch

Among other things, the challenges with low yields and perennial weeds in the clover ground cover have meant that in recent years the focus has shifted more toward the no-plow cultivation of cereals and other annual crops in mulch.

The method consists of cultivating strongly growing catch-crops, such as rye, that in spring is either cut down or flattened and cut into pieces with a special tractor implement called a roller crimper. This layer of composting plant matter on top of the soil suppresses the weeds. The roller crimper can be mounted on the front of a tractor, and a seed drill that is specially developed for sowing through a cover of straw is pulled behind it. The sowing can be completed in a single work step.

The method appears to be able to produce yields at the same level as organic cultivation with plowing.[43] Just as when growing grain in a living cover crop, you will be able to retain the nutrients because there is plant growth on the field almost all year round.

It is a cultivation method that requires precise timing, and when growing organically, the challenge is whether we can keep control of the weeds, and whether we can refrain from plowing completely over a longer period of time.

Perennial grain

Trials have been going on for many years to breed perennial cereal varieties by crossing cereals with perennial wild grasses. With perennial grain fields, we could use mechanized cultivation, where the same crop was harvested year after year without plowing, ideally in a polyculture with perennial beans and sunflowers, which are also being bred. Such a perennial polyculture would be the most ideal way of growing cereals in line with permaculture principles.

However, it has proven to be difficult to develop perennial grains, and the plant breeders in the area estimate that it will still take many years before we have cereal varieties that can grow for several years in a row and maintain an acceptable yield. It is therefore not yet a solution that is ready to be implemented in permaculture farming.

No-dig gardening with annuals

Annual beans, peas, root vegetables and onions complement the forest garden, as they can be stored for the winter months, when the forest garden does not produce many greens.

Digging, like plowing, destroys the soil's mycorrhiza and kills large parts of the microlife. For life in the soil, digging the garden is like causing an earthquake in a city of microorganisms, small animals and fungi. Therefore, annuals are grown in no-dig beds, in line with the permaculture principle of no-till.

These are permanently laid-out beds where we create a fertile, loose soil and only walk on stepping stones or paths between the beds.

Modest carbon storage from no-till cultivation

Forest gardens and agroforestry result in large carbon storage; carbon storage from no-till cultivation is relatively modest.[44,45]

Often, the increase in carbon in no-till beds occurs primarily by bringing carbon-containing materials such as manure, compost, cardboard, straw or dead leaves into the beds as an input from elsewhere, where the carbon level then becomes lower than it would otherwise be. In this situation, it is not the cultivated annuals that create the carbon storage.

Creating no-dig beds

If you are going to cultivate an area that has previously been a building site or a field that has been plowed and driven over by heavy machinery, the soil may be so compacted that even weeds do not grow well. Here it may be a good idea to loosen the soil with a fork before we start cultivating the soil.

With the double digging method, we can loosen the soil all the way down to a depth of 40cm so that the roots of the crops can grow unhindered in depth.

Emits CO_2 from over-oxygenation of the soil and machinery

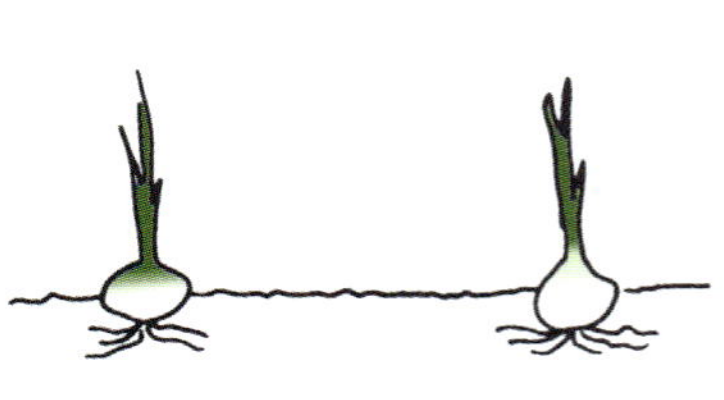

Fields that are tilled

Modest carbon storage

No-dig beds

Large carbon storage in soil and plants

Forest garden/food forest/agroforestry

Double digging is carried out by digging a trench to the depth of a spade in the desired width of the bed and putting the soil aside in a pile or in a wheelbarrow. The bottom of the trench is then loosened with a fork, and the next trench can be dug and the soil placed on top of the previous one. Continue in this way for the desired length of the bed, and finally fill in with the soil that was set aside. The soil has now been loosened once and for all to a depth of 30–40cm, and your no-dig bed has been created.

If the soil is not compacted, no digging should be done, and the beds can be created as lasagna cultivation, for example on a lawn, as described in the section on forest gardens (p.59).

The no-dig beds at Leif Varmark are grown in layers with apples over hokkaido squash over cabbage. Denmark.

Natural and loose soil

After the no-dig beds have been created, we maintain a healthy, loose soil that is rich in organic matter. Permanent paths or stepping stones are established to avoid compaction of the soil.

The healthy soil without compression damage allows plants to send their roots deeper to absorb water and nutrients, rather than only having to compete for nutrients in the upper approximately 20cm, as is the case in a field that is plowed. Therefore, the plants can be planted very close together, and the area where bare soil has to be prepared for a seedbed becomes smaller, which in itself is labor-saving. The area used for paths is more than compensated for by the close plant spacing.

The broadfork can be used to quickly loosen the soil at depth by double digging or to loosen the soil where a bed with lasagna cultivation is to be created. Calendula Permakultur, Denmark.

No-dig cultivation is intensive

No-dig beds belong in zones 1 and 2. Here, several crops are often grown one after the other in a growing season, and the yield per square meter is high. Therefore, no-dig beds usually require the addition of fertilizer and water, which are brought in from the other zones in the design.

Polycultures may be grown where crops with deep roots, such as carrots, are planted between onions, which have a superficial root system. Other beneficial relationships can then be considered, such as the smell of the onions masking the scent of carrots, which means that carrot flies are less likely to be attracted to the area. Or nitrogen-fixing plants, such as beans, are planted together with corn and pumpkins, which need a lot of nutrients. The last-mentioned polyculture is called the Three Sisters and is still used by the Indigenous population of Central and North America.

Each bed with its mono- or polyculture is preferably part of a crop rotation, where the different crops move to a new bed each year to avoid overwintering species-specific pests and diseases.

Pre-germination and planting

The yield per bed can be further optimized by pre-germinating the vegetables and planting them out. It is a trade-off, as it is a time-consuming extra step, but it can help us achieve more and earlier yields, and weeding is reduced as the soil is never bare.

The Three Sisters are a very productive polyculture, where pole beans, which are nitrogen-fixing, grow up tall corn plants, and plants from the pumpkin family keep the soil covered. The Three Sisters have been grown by the Indigenous people of Central and North America, and here it is seen at Nordhøj Permakultur, Denmark.

Design of the no-dig garden

The microclimate

Since no-dig gardening is an intensive cultivation system, it pays to put some effort into the construction of the beds and the microclimate around and on them.

If the garden or each bed is laid out with a slight slope of five degrees towards the south, the beds will warm up faster in the spring and the plants will, with the higher temperature, have greater photosynthesis, greater growth and a longer growing season.

The shape

In a small garden, we can work with keyhole beds, which give a larger area of bed per area of path.

Different garden designs.

Rectangular beds are easiest to handle in a larger garden. They make it easy to set up tunnel hoops with a cover of fleece or insect netting, and we can also have a chicken tractor that fits them. The beds are often made 75cm wide. With this width, it is possible to step across the bed from one path to the next, and tools have also been developed that fit this width.

In northern latitudes, where sun and heat are often the limiting factors for plant growth, the beds are placed south–north, so that all plants get either morning or afternoon sun.

Alternatively, on sloping land, the beds can be made along the contour lines so that they can absorb water that would otherwise run down the hill. On steeply sloping land we have to make terraces, which is such a big job that it is best to use another area for no-dig beds, if possible. In such a case, the *landscape profile* can have more influence than the zones.

Edges

The no-dig beds can be edged with boards, stones or similar, to prevent people from stepping on them. This may be necessary in public gardens. The edge can also provide an extra high bed that is comfortable to work with. However, high edges will drain the soil and are therefore mostly an advantage in wet areas, but a disadvantage in dry areas. The edges are

Bjarne Wickstrøm's no-dig garden is located in a clearing surrounded by trees. He uses polycultures and crop rotation. The paths are covered with wood chips from his homestead. Denmark.

The straight beds make it possible to use tunnels with fleece or, here, insect netting to protect the cabbage. Hjortespringgård Permaculture. Denmark.

a lot of work to create, and they will also provide hiding places for slugs, which is a disadvantage, especially for newly sprouted annual vegetables that can be completely eaten.

Nutrients in no-dig beds

Mulch

In no-dig beds, mulch is used, just like in nature's forest floor that is covered with leaves. This keeps the soil protected from drying out and from erosion, and at the same time we add nutrients and provide food for the living organisms in the soil. Earthworms continuously bring mulch into the soil, where it decays and nutrients become available to the plants, and the worm holes constantly loosen the bed; we let nature work for us. If organic matter without weed seeds is used, keeping the soil covered will also reduce weed germination.

Uncomposted organic matter such as leaves, hay or comfrey can be used for mulch and placed between the plants during the growing season. It is not practical to sow through such a cover, but we can plant pre-germinated plants through it. We can also spread the plant matter between plants that are already well spaced, but it is too time-consuming to place it between small, densely growing plants.

In spring and summer, easily degradable green material with a high nitrogen content is often used for mulch, e.g. comfrey, grass, alfalfa or seaweed.

In fall, if it is too late to sow a cover crop, the beds are completely covered with coarse material such as straw or leaves collected from a nearby forest.

Mature compost can also be used for mulch. If it is crumbly, it is possible to sow directly into it. Before sowing or planting, composted animal manure or nutrient-rich kitchen compost might be mixed into a cover of more nutrient-poor mulch.

Carbon loss and land footprint

Unfortunately, when we use mulch, part of the organic material comes into contact with oxygen and is lost as CO_2 to the air. Importing organic matter for mulch is therefore not a solution that mitigates climate change.

If we grow vegetables and other annual plants on a garden scale, we may not notice that cultivation through mulch requires high inputs of both labor and organic matter. But when we are going to grow, for example, 1000m^2 or more for sale, it is not realistic to collect tonnes of mulch material only to watch it decompose and know that a large part of it disappears into the air as CO_2.

In some cases, no-dig market gardening is based on importing several truckloads of compost or deep bedding each year, for mulching with.

It's important to be aware that although no-dig gardens are highly productive, they have a hidden land footprint that comes from the production of plant material and manure used as mulch. In many cases, composted manure from ruminants is used, making the garden dependent on farms with high methane emissions.

Green manure

We can grow our own material for mulch by planting beds with alfalfa and comfrey, both of which are long-lived perennial plants, next to the beds with annuals. Then we know how much area is occupied for this purpose and that it is cultivated in a responsible manner.

At Sieben Linden, a no-dig garden is grown that does not use livestock manure, but where both catch crops and green manure are part of the crop rotation. They call it a vegan garden. Germany.

We can also sow full-season green manures that are part of the crop rotation with annual vegetables. Examples show that by spending two years out of a seven-year crop rotation on green manure plants, it can be enough to provide the necessary nutrients without importing fertilizer.[46]

Catch crops

In reality, not very large amounts of nutrients are exported via the vegetables we can grow and sell. If we avoid losing nutrients from the beds, we do not need to import very much fertilizer. The most important way to avoid losing nutrients is to ensure that plants are constantly growing on the ground, either vegetables or catch crops.

This is in contrast to leaving your beds bare all winter, where there will be a large loss. Oilseed radish and winter rye are examples of catch crops. Catch crops must be sown at the end of the summer and cannot establish themselves after late-harvested crops.

The 3 hectare large garden supplies all kinds of fruit and vegetables to the Sieben Linden ecovillage in Germany, of 100 adults, 50 children and volunteers. Here you can see squash, corn and cold frames between avenues of fruit and berries.

Can we achieve forest soil in annual cultivation?

The manure normally used in organic farming is based on straw from cereals and grasses. Interestingly, organic matter from trees is thought to create more lasting humus than straw when added to the soil. Spreading wood chips, from mainly deciduous trees, on the field is one way in which we can bring the qualities of a forest soil into a growing system with annual plants.[47]

Iain Tolhurst of Tolhurst Organic farm in England doubled the yield of vegetables on his farm when he started adding a very thin layer of composted wood chips twice during the crop rotation. He believes that by doing this he has found a balance where he adds enough carbon to stimulate the fungi and bacteria in the soil but not so much that the carbon binds nitrogen in the soil, which is then unavailable to the crops.

Branches with bark on them can be chipped and used uncomposted in the same way as composted wood chips and without the loss of carbon associated with composting. Tolhurst is now planting trees on the farm so that in the future he can produce the carbon that is put on the beds himself.[48]

Liquid fertilizer

In addition to mulch and green manure for annual plants that require fertilizer, it is possible to supplement with diluted urine or with liquid plant manure, where, for example, nettles or comfrey plants are soaked in water and provide a nutrient-rich fertilizer juice. The liquid fertilizer is applied only during the growing season, when the plants have established roots and leaves and are really growing.

No-dig market gardening

In recent years, many small farms have started up with no-dig market gardening around the world. It is a way of starting to produce for sale in a short time and with relatively little investment. It only takes a small area to be able to create a full-time job with market gardening.

On Bec Hellouin farm in northern France, a research team investigated a cultivation area of 1000m^2 with 73 different crops, 40% of which were in polytunnels. The researchers found that the average work requirement was 43 hours per week, including time for marketing and accounting, and the income was at a level that was acceptable to farmers in the area.[49] It is the principle of *small intensive scale* with *high diversity* and *high yield* that becomes clear.

Special tools

Innovative market gardeners, such as the American Elliot Coleman, have developed new hand-held machines and tools specifically for no-till cultivation in 75cm wide beds. This has helped to make market gardening more competitive.

This includes:

- A small six-row seeder that is good at sowing small seeds, such as carrots and spinach, densely and precisely so that they quickly come up and cover the bed and do not need to be thinned.
- A broadfork makes it easier to loosen the soil without hurting your back.
- A mini tiller powered by a cordless screwdriver is suitable for making a fine bed.
- A salad harvester also powered by a cordless drill can make harvesting more efficient.
- A wheelbarrow with an electric motor makes the heavy work of transporting compost, vegetable boxes, etc. easier.

In recent years, there have been many examples of small-scale farms starting up with no-dig market gardening for local sales either with subscription customers, market stalls, farm shops or as sales at events. Often part of the production takes place in greenhouses.

On the family farm Calendula Permakultur, the market garden is just a part of their permaculture farm which also includes fruit, berries, chickens, ducks and sheep.

MATHIAS OLSEN

Palle's garden farm includes vegetable production, animals and a farm shop, and is open to visitors.

NIKLAS MILTHERS

Nordhøj Permakultur in Denmark brings vegetable boxes out on an electric cargo bike. Here, the market garden is also part of a larger permaculture design, including grapes for wine production.

No-dig beds with biochar-mulch at Edward Revill's, in Wales.

EDWARD REVILL

Biochar and terra preta

Biochar and microbial carbonization may be the pieces we are missing in permaculture to create a cultivation system with annual crops that counteracts climate change. Here we describe how biochar can be used for cultivation. In the 'energy technologies' section (p.157), we discuss how we can produce biochar by burning wood in different types of stoves.

Cultivation with biochar originates from the Amazon rainforests from before colonization. Under normal conditions, the decomposition of organic matter and therefore carbon in the soil is rapid in the warm, humid climate of the Amazon, so even though large amounts of organic matter continuously fall to the forest floor, the soil layer is only about 10cm deep. Where people in the past added biochar together with organic material, the soil layer could be two meters deep. A particularly interesting observation is that it is not only the biochar part of about 9% that has remained in the soil; much of the organic material, such as fish bones and animal and human feces, can still be identified.

These man-made soils are called terra preta, which means black soil in Portuguese, and they are still fertile and nutritious, even though they have existed since the time of Columbus. This is because biochar has an enormous surface area and at the same time a strong electrostatic charge. The electrostatic charge causes organic material that would otherwise have been decomposed or washed away to stick to the surface of the biochar. One theory is that biochar inhibits the normal decomposition of organic material by bacteria and fungi.

Today, we have lost knowledge of how exactly terra preta was produced, but it is clear that the soils have functioned as a stable carbon store for several thousand years. Therefore, there is great interest in the permaculture movement to imitate this terra preta property and turn the CO_2 emissions from the cultivation of annual vegetables and grains into carbon storage that counteracts climate change.

EDWARD REVILL

When harvesting potatoes, you can see how black the soil has become.

Mulch with biochar

The potential for improving the soil with biochar is greatest if we have sandy infertile soil as a starting point. Here, biochar can make a significant difference and improve the soil's ability to retain water and nutrients.

Most focus has been on using biochar in such a way that it is incorporated into the soil by plowing. But if we want to use biochar in no-till systems, the most interesting thing is probably how we can make mulch that includes biochar.

EDWARD REVILL

On Revill's lettuce roots, the mycorrhiza fungi are visible as white threads. They increase carbon storage enormously and make more nutrients available to the plants.

This has been tested by Ed Revill in Wales.[50, 51] Biochar is crushed into small pieces and mixed with other organic matter, for example in the bedding under his animals and in the compost pile. During the composting process, strong chemical compounds were formed. By mixing it with compost, biochar becomes heavier so that it is not lost to water and wind erosion.

The compost is used as mulch on the no-dig beds in an alley-cropping system with annual vegetables and grains between rows of trees. The trees are kept small by pruning and coppicing to avoid too much shade for the sun-demanding annual crops, and the wood prunings are used to produce biochar.

The experience is that the mulch with biochar, instead of being decomposed over the summer, as normal 'fresh' mulch does, becomes a stable black and nutrient-rich surface layer with a good crumb structure, which is easy to cultivate in. During the growing season, this layer is supplemented with liquid fertilizer or fresh greens.

The mulch with biochar seems to have several advantages:

- Due to the biochar's large water-storage capacity, the system becomes more resistant to drought.
- Because the mulch is more stable, less material has to be imported from outside to cover the soil with.
- Carbon is stored so that the system as a whole mitigates climate change.

Compost

Compost is an important part of establishing closed nutrient cycles in gardens, on farms and in cities. This is because composting allows us to convert organic waste into a soil-improving fertilizer.

Nitrogen is bound in compost, which means that it is converted to solid form so that it is released slowly to the plants and we avoid leaching and evaporation. One goal of composting is also to preserve the carbon in the plant matter in a permanent form as humus so that it is not quickly lost from the soil as CO_2.

The graph shows how different forms of fertilizer release nitrogen to the plants. It appears that compost is the form of fertilizer that releases the most to the plants at the end of the summer when the plants need it most and again the following year.

In fact, the compost will continue to be of benefit in the third year and beyond. Compost is thus a way of building up fertile soil, while liquid fertilizer only delivers nitrogen to the plants in the months immediately after application.

Hot compost

A hot compost is characterized by the fact that it reaches a temperature of 60°C or more during the first week after setting up. A hot compost is made by collecting a minimum of 1m^3 of fresh material in a pile in a compact shape.

The materials must be combined so that there is a reasonable ratio between carbon (C) and nitrogen (N) of approximately 25:1. This ratio is present, for example, in horse manure without straw added, while in, for example, wheat straw there is a C:N ratio of 128:1, and in fresh grass the C:N ratio is 12:1. The materials are mixed evenly, and water is added so that they feel moist but not wet.

During hot composting, nitrogen can be lost through evaporation and seepage,

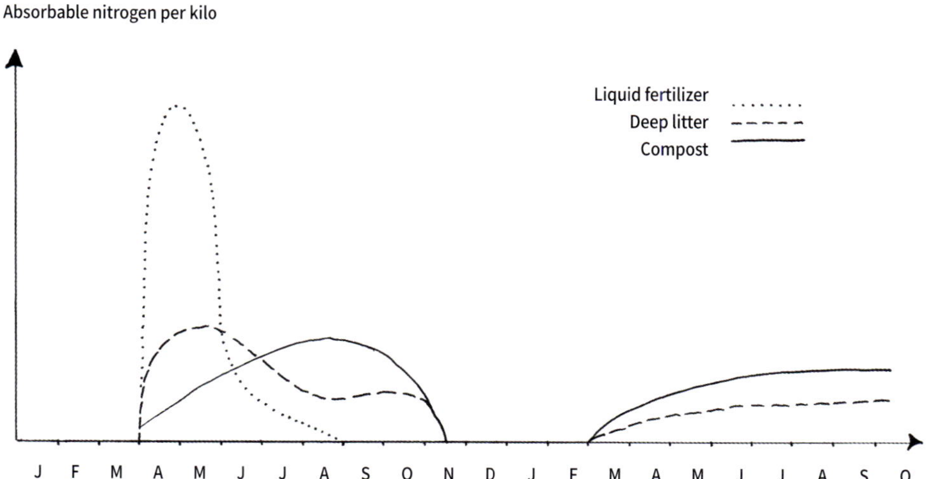

To make hot compost, there must be a minimum of 1m³ of unconverted material. Here, grass clippings are mixed with deep litter from the goat barn and watered for the right level of moisture. The pile is covered, and after a while it is turned over so that the outermost material is moved to the inside. The high temperatures kill weed seeds and pathogens. Reforest Farm, Denmark.

therefore the pile must be covered with a 'compost fleece' or a tarp. A compost fleece is good because it sheds rainwater but is 100% breathable. If a plastic tarp is used, it is important that it does not reach all the way to the ground, so that air can enter the composting process. The pile should have a layer of straw or similar at the bottom to absorb the water with nutrients that seeps out of the pile. Drainage pipes can also be placed at the bottom to improve oxygen access to the center of the pile. When composting in the winter, the pile should especially be insulated on the outside with a layer of straw or similar.

The microorganisms will work so hard that the temperature increases. When the temperature in the pile begins to drop, the compost is turned over so that the outermost part of the pile is on the inside and exposed to the high temperatures. After a few months, the compost is so well composted that we can start using it as mulch.

The high temperature during hot composting kills weed seeds and disease germs. Therefore, the finished hot compost is ideal to use in an annual no-dig garden or to establish new beds with perennials in the forest garden. A *beneficial relationship* can be made by making your hot compost in the greenhouse in the spring because the plants can benefit from the heat, so it will be possible to get an earlier start to the growing season.

Microbial carbonization

A composting method that is particularly effective in converting lignin (the chief constituent of wood) into durable humus is microbial carbonization.[52] The method differs from ordinary hot composting in that a higher proportion of carbon can be included in the mixture and that the pile is not turned over.

The following materials are mixed well:

- 40–50% straw, wood chips, bark, garden waste or reeds
- 30% maximum grass, clover or fresh green waste
- 20% or more solid manure or slurry from cattle, pigs, sheep, poultry or horses
- 5–10% clay and mature compost – the mature compost helps to steer the microbial processes in a positive direction
- It is possible to mix in biochar, 1kg of lime and/or 3–4kg of rockdust per ton.

The substrates must not be rotten or have fungal growth. If necessary, water is added to the materials while they are being mixed so that they feel moist but not wet. Dry materials such as straw and wood chips can be difficult to moisten and should be soaked a few days in advance.

The pile is made trapezoidal (flat on top) and between 1 and 2 meters high. The pile is pressed relatively firmly together, as some of the processes desired are anaerobic (oxygen-free). Finally it is covered with a layer of soil. In some countries there is a requirement to cover the pile with plastic when it contains livestock manure. This is problematic since the surface should have as much light and air as possible. A compromise is to cover the pile with a thin fleece that lets light through. Composting is usually finished after 8–12 weeks, but if wood chips are included, the process will take over half a year.

HELGE BÜLOW

A power plant built from wood chips, vegetable waste and chicken manure, which will heat one of the greenhouses at Gerlev Idrætshøjskole, Denmark.

Vermicompost

In an ordinary household, it is difficult to collect sufficient amounts of material all at once to make a hot compost or microbial carbonization. Here it is more practical to use a rat-proof composter that we fill continuously. The compostable household waste that we have available in an ordinary household, especially kitchen waste, often has too little carbon in relation to nitrogen to be composted without the addition of straw or wood chips. The kitchen waste will instead clump together and rot. Here, vermicompost is a solution.

If we add composting worms (*Eisenia fetida)* to a household compost, the worms can convert the material into compost in a short time. However, vermicompost can also become too wet, so we have to add some straw. Vermicompost has the advantage that it does not take up much space, so we can use it in densely built-up areas. It can even be made in a closed container with air holes, in an apartment.

The vermicompost must be added with appropriate amounts of new material approximately every other day to keep the worms fed; it should be covered to avoid flies and the like. The worms thrive best at 23°C and die at 30°C. At low temperatures, the worms convert less material, and in frost they move downwards in the compost and clump together.

Composting worms cannot burrow into the soil like earthworms but mainly move through looser material. Therefore, a 'worm tower' – a vertically buried tube with composted material in it and with holes on the side – can be of help to the worms. They can then travel through different temperature zones and bring organic material out through the holes. In case of prolonged frost, we can take some worms into the house to ensure that they survive. The finished compost is nutritious and can be used in the no-dig beds.

Compost worms exist sporadically in nature but can be purchased in nurseries to start a compost.

BIRGIT ROTHMANN

Fresh kitchen waste is added to the worm tower every other day so that the worms are fed. Birgit Rothmann has made the towers in the middle of her raised beds so that they are fertilized. Denmark.

Nutrients in forest gardens and agroforestry

When many new forest gardens, food forests and agroforestry systems are designed and planted in the future, it is important that the plants have the nutrients available that they need to give their optimal yield. Only in this way will we create the highly productive alternatives to industrial agriculture that are necessary in a densely populated world.

In this section, we describe how we obtain nutrients for cultivation, and how we calculate whether the plants' nutritional needs are covered in our permaculture design.

The nutrient account is made while we are designing and before we plant the canopy layer. By spending some time on this upfront, we can create a nutrient cycle that will work for a long time into the future.

In some European countries, there is a legislative requirement for farmers to have fertilizer accounts, where it takes time and often consultants to work out the correct amount of fertilizer to be used on the land.

Despite this, there are problems with the leaching of nutrients into groundwater, lakes and streams or evaporation of nitrogen into the air.

Doing nutrient accounting in a permaculture growing system thus makes it clear how fantastically simple a permaculture system can make something that is complex and inappropriate in today's agriculture.

As a general rule of thumb, the leaves of the nitrogen-fixing trees on the left are green and rich in nutrients when the leaves fall.

Left column (top to bottom):
Italian alder
Siberian pea shrub
Black locust
Sea buckthorn
Right column:
Pear
Beech
Sweet chestnut

The design map for Reforest Farm shows the nitrogen-fixing trees and shrubs in red.

It becomes simple because we work *with* nature instead of *against* it. We plant perennial plants and let nature create a well-developed root system, where we get phosphorus from the soil using a network of fungal roots (mycorrhiza), potassium from the soil with accumulator plants and nitrogen from the air with nitrogen-fixing plants. When we then, at the same time, give back our own fertilizer (recycle the products from a compost toilet and wastewater) we can create resilient growing systems that are self-sufficient in nutrients year after year.

Nitrogen, phosphorus and potassium are the three primary nutrients, which is why these three nutrients are what we are exploring here.

Nitrogen from trees

In conventional agriculture, nitrogen for fertilizer comes from factories that make nitrogen from the air in solid form. However, this process is very energy intensive, and 41% of the energy consumption for conventional crop production can be attributed to nitrogen fertilizer.[53]

Nitrogen for the cultivation system can also be obtained by using nitrogen-fixing plants which use only energy from the sun. Just like in factories, the nitrogen-fixing process requires a lot of energy, and in nature this means direct sunlight. Nitrogen fixation in an agroforestry system is therefore not as efficient in the shade under the trees, and the main part of the fixation must happen in the upper canopy layer and must come from nitrogen-fixing trees and also shrubs between the trees.

The nitrogen-fixing trees and shrubs distribute their nitrogen in three ways:

- **The leaves** that fall from the nitrogen-fixing trees contain nitrogen, which will become available to other plants through natural turnover in the forest floor.
- **Fine roots** from nitrogen-fixing trees and shrubs will die off each year, releasing nitrogen. If we cut branches from nitrogen-fixing trees, more roots will die off at this time, and a particularly large amount of nitrogen will be released.
- **Mycorrhiza** distributes nitrogen from areas with high nitrogen levels, such as those found in nitrogen-fixing plants, to areas with low nitrogen levels.

These three methods all present no risk of evaporation or leaching, as permaculture farming works without plowing, and there is always a well-developed root system from perennial plants that can absorb the available nitrogen.

Nitrogen-fixers with multiple functions

In permaculture design, we always try to ensure that each element has *multiple functions,* and thus we can choose nitrogen-fixing trees and shrubs that also provide shelter, berries, edible seeds, timber, firewood and/or leaves for fodder.

Potassium from accumulator plants

Potassium is mined in many places in the world, and the largest deposits are found in Canada, Russia, Belarus and Germany. It is a nonrenewable resource, but it is found in large quantities. There is potassium in the cultivated soil, which various plants with deep roots will take up – the so-called accumulator plants.

Russian comfrey of the hybrid variety Bocking 14 is used in particular. It is good at taking up potassium and is also sterile, so it does not spread via seeds. The large soft leaves can be cut off three times a year and used as nutritious mulch in the forest garden or kitchen garden or as an addition to the compost. The plant is propagated by dividing the roots and planted in new patches in forest gardens, no-dig gardens and chicken coops.

Most accumulator plants accumulate several different nutrients and minerals for the benefit of the system. With great diversity in our permaculture cultivation, we therefore have them all covered.

In addition to comfrey, many other trees, shrubs and ground cover plants are also accumulator plants, and this ensures the resilience of the overall nutrient system.

Examples of nitrogen-fixers

Ebbing's silverberry *(Elaeagnus × ebbingei) is evergreen, so the shrub provides shelter all year round. The leaves are eaten by chickens, so it is an obvious choice in the winter chicken forest. It flowers in October and is therefore a good bee plant. Over the winter, it stands with green berries, which are harvested in the spring before other berries are available. They are rich in essential fatty acids, vitamins, minerals and bioactive substances.*[54] *Reforest Farm, Denmark.*

Autumn olive *(Elaeagnus umbellata) and* ***Goumi*** *(Elaeagnus multiflora) produce tasty healthy berries in fall. There are high-yielding varieties with different harvest times. The shrubs also fix nitrogen in partial shade. Gammelgaard, Sweden.*

Italian alder *(Alnus cordata) in the middle, loses its green, nutrient rich leaves late. It is so tall it can reach the sunlight between tall nut trees. It has a narrow growth habit and therefore doesn't cast much shadow. It produces protein-rich leaf fodder. The species is more drought tolerant than common alder. Reforest Farm. Svanholm.*

Common alder *(Alnus glutinosa) produces protein-rich leaf fodder, and because it is native to Europe, it promotes insect life. Common alder can grow in very wet soil.*

Nitrogen fixation from the alder requires full sunlight and occurs in the layer above the 'lettuce tree' lime. Martin Crawford's forest garden, England.

Robinia *(Robinia pseudoacasia) has a trunk that can be used instead of tanalized/preserved timber and flower clusters that can be eaten. The canopy only casts light shade, so it can be underplanted with shade-tolerant species. Reforest Farm, Denmark*

Siberian pea shrub *(Caragana arborescens) sheds its seeds with high protein content, and is therefore an obvious fodder plant in the chicken forest. The seeds are also edible by humans, but they take a long time to harvest. Reforest Farm, Denmark.*

Sea buckthorn *(Elaeagnus rhamnoides formerly Hippophae rhamnoides) is wind-resistant and produces sour berries with a high vitamin C content, which are used for wine and juice. The berries can also be used as oily chicken feed, which the chickens harvest themselves. High-yielding sweet varieties have been bred. However, it can become invasive when introduced to non-native landscapes, so this needs to be considered. Reforest Farm, Denmark.*

The accumulator plant Russian comfrey of the hybrid variety Bocking 14 has purple flowers that are beneficial to insects. Here it is also used as a barrier between two beds on Reforest Farm, Denmark.

Other accumulator plants include

- Perennial vegetables such as sweet cicely, fennel, liquorice (fixes N), all kinds of sorrel and nettle.
- Plants that also benefit bees and other insects such as dandelion, lupin (fixes N), vetch (fixes N) and yarrow.
- Trees such as lime, apple, walnut, robinia (fixes N) and common alder (fixes N).[55]

Phosphorus with help from mycorrhizae

There is plenty of phosphorus in the soil, but it is difficult for plants to access. In conventional agriculture, phosphorus is therefore extracted from mines and is a nonrenewable resource. This primarily concerns mines in Western Sahara, which are occupied by Morocco, which has displaced the Indigenous population in order to gain control of the mines.

Phosphorus is a resource of which only small amounts remain; in the future this could cause major problems in relation to the world's overall food supply. With forest gardens and agroforestry, we ensure the supply of phosphorus entirely through nature's own process. Mycorrhizal fungi are always present in the soil, but they are set back when they are cut and turned during plowing and other cultivation of the soil. That is why there is little mycorrhiza in agricultural soils today.

Conversely, when the soil in permaculture systems is not disturbed, the mycorrhizal fungi in the soil will develop into a large underground network that can extend for kilometers. The fungal filaments form a thin slimy layer around the roots and send out long filaments from here. The fungi form a symbiosis with the plant, where the fungus, which cannot photosynthesize, receives sugar from the plant and in return supplies the plant with water and nutrients.

The mycorrhizal fungus can thus make the plant's root network 80 times larger; it can reach into very small pores in the soil where the plant itself cannot reach and thus obtain the phosphorus that is otherwise not available to the plant. Phosphorus will always be present and available through natural weathering in the soil, as long as we do not disturb the mycorrhizal fungi.

Mycorrhizae can be accelerated

To accelerate the process of developing mycorrhiza, we can supply the trees we plant with mycorrhizae. We can buy a spore mixture and scatter it over the roots of the trees when planting, or we can collect mushrooms that we know are mycorrhizal fungi and dissolve them in water, which we can dip the tree roots in so that one mushroom can reach many trees. By inoculating with edible mycorrhizal fungi, we can also increase the likelihood that we will get mushrooms as an extra yield from our permaculture system.

ESBEN SCHULTZ

The tree roots are sprinkled with charcoal inoculated with fungal spores to accelerate mycorrhiza. Reforest Farm, Denmark.

Calculate nutrient requirements

The fungi ensure that the plants we grow are supplied with phosphorus, so there is no need to calculate the nutrient requirements here. It is necessary in relation to nitrogen and potassium, however.

Forest gardens and agroforestry systems usually have high diversity, with many different plant species in polyculture. When we need to find the nutrient requirement for these systems, it is therefore necessary to simplify and divide the plants into groups according to their needs. This method comes from the book *Creating a Forest Garden* by Martin Crawford. We have expanded it to also include people and animals and have adjusted the figures slightly,[56] and we suggest that it be used on permaculture systems more generally.

We calculate the area of the crowns of the fully grown trees and shrubs, with high and moderate nutrient requirements respectively, as well as the area of annuals with extremely high nutrient requirements (Table 1). Each area is then multiplied by the current need for nitrogen and potassium.

The need for nutrients that we have now found is based on a situation where we harvest the crops from a forest garden that we would normally harvest for people, i.e. fruit, nuts, vegetables and the like, and where we do not deliver the nutrients back.

If we harvest anything other than these

forest garden products, we take extra nutrients out of the system, and this must be added to the nutrient requirement. For example, we may have laying hens that find their food between the trees in the form of insects and seeds. In that case, the nutrients in the eggs must be added to the total requirement in whole or in part – we 'harvest' the eggs from the system.

Another example may be that we have animals for meat production that are fed with grass or leaves from the trees. In this case, the nutrients in the meat must be added to the total nutrient requirement. If the manure from the animals remains in the system, we can disregard the nutrients in the manure, but if it is removed from the overall system, the nutrients in the manure must be added to the total requirement.

If we purchase feed for our animals, e.g. grain, this must be included as an input of nutrients. It can be helpful to create an overview by making a circuit diagram with arrows that show the path of nutrients around the permaculture system.

How are the nutrient requirements met?

Once we know the total need for nitrogen and potassium, we can cover it in different ways. Nitrogen-fixing trees and shrubs will usually be the most important source of nitrogen. Here, the area of the crowns of the nitrogen-fixing trees is calculated. Each square meter of nitrogen-fixing plants provides 10 grams of nitrogen in full sunlight and 5 grams if they are in partial shade. However, nitrogen can also be obtained from other sources, see Table 2.

Table 1
Nitrogen & potassium requirements

Plants with extremely high nutrient requirements of 18g N/m²/year and 20g K/m²/year:
Cereals and annual vegetables (vegetables like carrots, spinach, peas and beans have a lower nutrient requirement).

Plants with high nutrient requirements of 8g N/m²/year and 10g K/m²/ha/year:
Bamboo, chestnut, hazelnut, walnut, apple, mulberry, apricot, plum, peach, pear, blackcurrant, gooseberry, blackberry, high-yielding perennial vegetables.

Plants with moderate nutrient requirements of 2g N/m²/year and 3g K/m²/year:
Cornelian cherry, juneberries, edible hawthorn, snowbell tree, sour and sweet cherry, elder, rowans, Sichuan pepper, raspberry, currant.

Plants that do not need to be supplied with extra nutrients:
Hazel for staking, fig, pine for pine nuts, juniper, trees and shrubs where the leaves are harvested.

Animal products:
Eggs weigh about 60g/each and contain 1.3g K and 20g N per kg. Meat contains about 4g K and 32g N per kg.

Table 2
Nutrient supply

Figures for calculating nutrient supply	N kg/year	K kg/year
Nitrogen-fixing plant full sun per m²	0.01	-
Nitrogen-fixing plant partial shade per m²	0.005	-
Comfrey plant annually with 3 prunings	0.0015	0.03
Urine from 1 person[a, b]	4.0	0.7
Feces from 1 person[c]	0.9	0.4
Ash from burning 1 tonne of wood[d]	0.1	2
Seaweed 1 tonne dry weight[e]	<12	28–100
Grain (wheat, barley, oats)[f]	24g N/kg	5g K/kg
Straw (wheat, barley, oats)[g]	13g N/kg	14g K/kg

a https://sswm.info/factsheet/urine-fertilisation-%28small-scale%29
b Harper, Peter, and Louise Halestrap: *Lifting the Lid, An ecological approach to toilet systems*, CAT publications, 1999, p.120.
c Harper and Halestrap: *Lifting the Lid*
d When burned, approximately 8% ash is obtained containing 2.6% K and 0.15% N (source: http://extension.uga.edu/publications/detail.html?number=B1142).
e Seaweed contains in relation to its dry weight. 1.2% N, 0.2-1.3% P, and 2.8-10% K (source: https://garden.org/learn/articles/view/379).
f www.lakeheadu.ca/sites/default/files/uploads/3470/Documents/Extension_Articles/Nutrients_removal_by_field_crops.pdf
g www.lakeheadu.ca/sites/default/files/uploads/3470/Documents/Extension_Articles/Nutrients_removal_by_field_crops.pdf

Forests as carbon storage and for local production

Forest management today

Since permaculture uses the primeval forest as a model, one can naturally ask the question of whether permaculture has any relevance in places where there already is a forest. However, if we look more closely at modern forestry methods, we will see that the permaculture approach is highly relevant in forests.

Most temperate forests today are very different from the primeval forest. They are managed on the clear-felling system, also called rotation management. In many places forestry includes operations such as site preparation like stub removal and breaking up soil by fracturing or tilling it, tree planting (including, sometimes, the use of genetically improved trees and/or exotic tree species), tending, thinning, fertilizer application and clear-cutting of sometimes large areas.[57]

Forests and climate impact

The old forests that still exist can store carbon for centuries and thus play a significant role in climate mitigation.[58] Additionally, old forests are important for biodiversity. They belong to the most species-rich ecosystems and should be completely protected against felling. The forest products that we need should come from forests which are already influenced by humans.

In the 5 ha. Gurre Skovhave farm in Denmark, the starting point was a dense old spruce plantation, which has now been thinned so that a few spruces remain and create a light open forest environment. Here, Pia Arentsen Willumsen has planted groups of pears, apples, aronia, currants, etc.

By leaving some trunks 2–3 meters high, a trellis for climbing plants has been created, here gooseberry kiwi. Gurre Skovhave, Denmark.

The carbon stores of our managed forests are affected by forest management practices and how the harvested trees are used.

As mentioned under the vision of reforestation of the earth, a large part of the carbon storage in a forest soil occurs when the trees supply sugars to the mycorrhizal fungi in the soil, and the fungi then store carbon from the sugar in the soil. This storage becomes most effective when the forest has been standing for a number of years so that the fungal network is well developed.

However, carbon storage is interrupted when the trees are felled. Instead, the forest floor releases CO_2 when all the trees have been felled and sunlight hits the forest floor, which then becomes unnaturally warm. For the first 10–15 years after a clear-cut, the forest will emit more greenhouse gases than it absorbs, and it will take 20–40 years until the emissions associated with the clear-cut are compensated for by the uptake from the new trees.[59]

The timber produced from a forest consists of carbon and can constitute a long-term carbon store if the timber is used in buildings, for example. Unfortunately, today's practice is that a very large part of the wood produced is used for purposes with a short lifespan, especially biomass for fuel and paper. It is only possible to achieve longer lasting products if the trees are of good quality in terms of age and the absence of damage from insects or disease damage. Forests that are closer to nature are more resilient and better at withstanding fires, pests and diseases. Furthermore, clear-cutting forests also increases the risk of storm damage in stands surrounding clear-cuts.

Lack of biodiversity and local production

In addition to the climate perspective, the conditions for biodiversity also speak against forest management with monoculture and clear-cutting because these provide poor conditions for the natural flora and fauna. If a large area is clear-cut, the habitat also changes over a large area, making it difficult for wild plants and animals to find suitable new habitat.

The wood that is felled is sold on the global market and must be transported over long distances, while the forested regions of the world often do not have their own food production and therefore have to import their food from the global market.

Permaculture and forestry

An approach to forestry that is more in line with permaculture than the clear-felling system is continuous cover forestry. This can be selective felling, where only individual trees are felled, or group selection of trees in small areas of up to 0.5 hectare so that the natural ecosystems can still be connected.

Continuous cover forestry most often relies on natural regeneration although trees are sometimes planted.

Most research is done on rotation-managed forests, so selection management needs to be studied more. But there is research indicating that selection-managed forests constitute a larger net carbon sink in the long term.[60, 61]

If a permaculture project has a forest monoculture as its starting point, we must prevent clear-cutting and instead develop it towards continuous cover forestry. In this way, we ensure that the carbon that is already stored is kept out of the atmosphere, and we promote conditions for biodiversity. An appropriate felling pace may be to fell only 20% of the trees every 20 years,[62] but of course it depends on the specific location, species composition, etc.

By switching to such a felling cycle, the forest as a whole will have more old and large trees, and the amount of bound carbon will increase. When we only fell the oldest trees, there is an undergrowth ready, which will grow vigorously and quickly. Therefore, such a felling cycle is also the one that gives the greatest growth overall.

By developing a forest toward continuous cover forestry, the forest will become more similar to a natural forest. Regeneration is controlled by selective cutting of trees, so the number of species is increased and so that there are trees of different ages growing close to each other.

As a general rule, we should only bring native species or at least species from the same continent into an existing forest, since this is basically zone 4 where biodiversity must have a high priority. An exception to using only native species may be if the existing forest does not initially contain a high biodiversity and is also located immediately adjacent to our home; we may therefore want to change it into a zone 1–3 forest garden over time.

For private forest owners, a decision to avoid clear-cutting and to let the trees grow older before felling them may mean that income from the forest

In an area with newly planted fruit trees, egg-laying hens have been introduced. The products are sold in local shops. Gurre Skovhave, Denmark.

is delayed. A way to get an earlier income from the forest could be to find other income opportunities than selling wood. One option is forest farming.

Forest farming

It is possible to produce food in a forest that is not initially designed as an agroforestry system, and here are some ideas for this.[63]

Wood that is not of timber quality can be used to grow edible mushrooms, sap can be tapped from birch and maple trees and possibly syrup can be produced. Shade-tolerant perennial vegetables can be grown on the forest floor, such as ramsons and ostrich fern, whose spring shoots taste mildly like artichokes when boiled. Linden flowers and other plants can be picked and dried for tea.

On the forest edges, where there is sufficient light, fruit, nuts and berries can be grown. Hazel and bladdernut are obvious options because they can produce nuts in partial shade. Elderberries, raspberries,

PETER COX

Peter and Marie Cox bought a 1.6-hectare plot of land, most of which was a monoculture spruce plantation. Here they have made a clearing, but left a few trees standing to preserve the forest environment, and then established 'The Self-Sufficient Forest Garden'. Denmark.

blackberries, gooseberries and currants are all well-known berries that would thrive here, while lesser-known ones such as aronia can also produce berries in a forest edge.

Animals can be integrated into the forest in a way that they exploit resources that would otherwise not be used, for example pigs and ducks can find much of their food in the forest themselves, and goats can be used to clear unwanted vegetation. It is of course important that animals are integrated in ways that do not destroy existing natural values.

Forest farming can give the forest owner an economic alternative to modern forestry and thereby help to secure the forest. Producing food in the forests also reduces the total need for agricultural land and thus the pressure to obtain more agricultural land through deforestation.

At the permaculture farm Alvastien in Norway, most of the area is forest. Here, they cut down linden trees and make linden bast, which is used for locally sustainable rope and wickerwork. Linden bast is the underbark of linden trees and is very strong. It has been used for millennia, among other things, for ropes for Viking ships, instead of today's nylon and imported plant fibers.

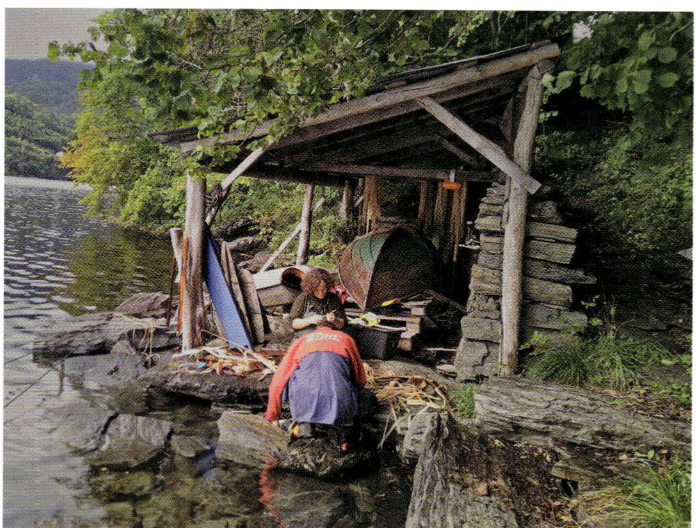

The inner bark of the linden tree must be soaked in water for 4–10 weeks, dried for a few weeks and then separated into paper-thin long pieces of bast. Hege Iren Aasdal from Alvastien rinses the linden bark in the local river.

THIS PAGE ALVASTIEN

Bag made of linden bast from Alvastien.

Land use

A whole country converted to permaculture

As described earlier, permaculture offers long-term solutions to a wide range of problems that global society is struggling with: the climate and biodiversity crisis, groundwater pollution, soil erosion and food insecurity, etc.

Therefore, permaculture methods need to be spread over large areas and designs, and implementation plans must be made for converting entire countries to permaculture. As an example of this, over the next few pages we will present a proposal for how the land distribution in Denmark could be changed in line with permaculture.

Today, Denmark is characterized by large monocultures with annual crops, which are primarily used as feed for a large animal production from cattle and pigs. Second only to Bangladesh, Denmark has the largest share of land which is being plowed.[64]

The area proposal shows how we can create a mosaic landscape, where the land is mainly reforested by different forest types, and only a small part is used for fields and vegetable gardens. Permaculture systems, such as food forests, chicken forests and nut forests, will grow and contain far more biodiversity than agriculture today and at the same time store carbon.

The example helps to emphasize that if we as a global society choose to convert our agricultural systems to permaculture farming, we will actually be able to tackle the biodiversity and climate crisis, which for many people may otherwise seem unmanageable or even inevitable disasters. The first step toward change is that we can see the vision before us. So what will the map look like if we convert agriculture in an entire country to permaculture?

Area types divided into zones

We have defined 11 different area types which we have categorized within the five zones in permaculture, and together these contain all the different permaculture cultivation systems described in this book.

Area distribution depends on the country

When we review the different area types, we come up with a suggestion for what percentage each area should cover in Denmark. See Table 3 on p.127.

What area should be covered by the different area types depends on which country we're working with. In countries like Denmark, which have a high population density and good arable land, zones 0–3 should take up a lot of land, and there will be less land for zones 4 and 5 in percentage terms.

Many other countries have larger areas that are sparsely populated, because there are mountains or deserts, for example, or areas that are too cold to farm. Here there will naturally be room for more zones 4 and 5.

Plant-based forest diet

An important prerequisite for the area proposal is that we must change our eating habits to a primarily plant-based diet because this reduces the need for agricultural land and can give as much land as possible back to nature.

An analysis of the global food system has shown that if we switch to a diet based on plants supplemented with a small amount of fish, eggs and meat from poultry, i.e. without milk and meat from ruminants, we will be able to reduce our land use on a global scale to a quarter of what is currently used for food production.[65]

According to the analysis, it will not reduce the land requirement much more if we also refrain from eating fish, poultry and eggs. This conclusion seems logical because poultry is easy to integrate into permaculture systems, where they can, for example, make use of otherwise untapped resources under the trees in the form of insects and greens for feed.

In particular, ruminants are minimized

We propose reducing the number of ruminants to one-sixth of the current number. This should be seen in relation to the fact that today there are six times as many large ruminants on the planet as there were 500 years ago. The remaining ruminants would be used for nature conservation in meadows, silvopastures and natural forests. Their primary task would be to promote biodiversity, and not to produce dairy products and meat. A reduction in the number of ruminants would make an important contribution to a rapid decrease in the content of methane in the atmosphere, which could help prevent us from exceeding climate tipping points. See the section on ruminants.

Animals in self-feeding forests

Other animals, such as poultry and pigs, do not emit methane to any significant extent and can therefore be included as an extra layer in food-producing forests, where they have a useful function for the cultivation system. Here, native breeds are used that are more frugal, and so they can get a large part of their feed from the forests themselves, meaning that there is no need for fields for animal feed. This will result in a smaller production of eggs and meat that can supplement a plant-based diet.

Food from sources other than agriculture

When we have to assess how much land we need for agriculture, it is also important to assess how much food comes from sources other than agriculture. This applies to private gardens, fjords and the sea, forests and open habitats. We give conservative figures for this, and it will probably be possible to obtain larger amounts of food from these areas.

Feeding 12 million in Denmark

There are six million people living in Denmark, but in the example, we assume that the country must produce food for 12 million people, as that is roughly how many people Danish agriculture can feed today.[66] In doing so, we take into account that there are other countries where people are dependent on importing food.

How are the areas calculated?

The calculations are based on the fact that a person should have 853,000 kcal/year and on how many calories the different types of land produce per hectare. It is of course a simplification to only calculate our diet in calories, but this allows for a comparison of the cultivation systems.

New land use –
Permaculture reforests the Earth

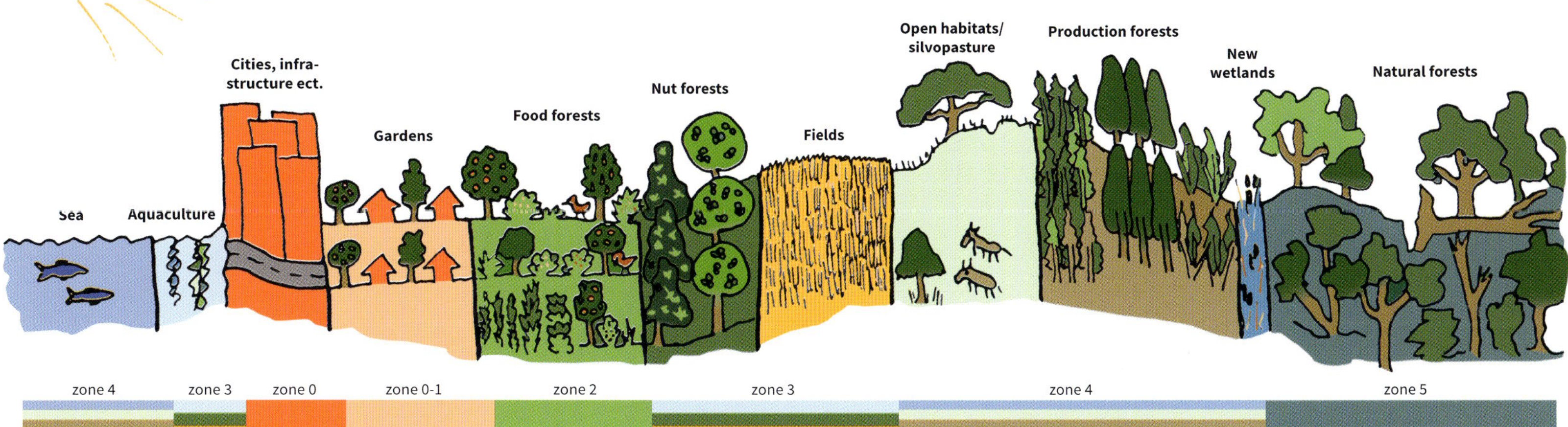

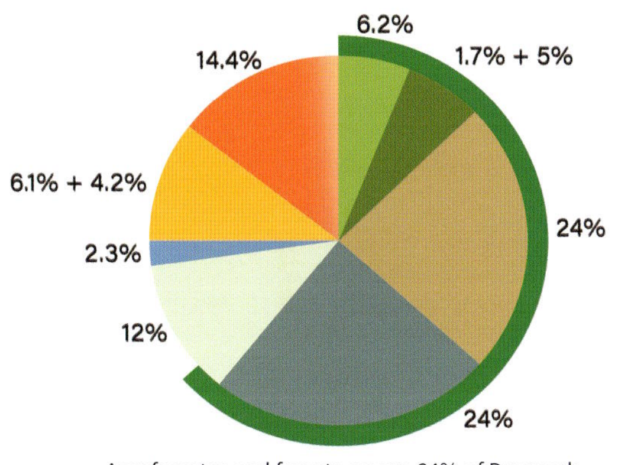

Agroforestry and forests covers 64% of Denmark.

About the drawing:

The drawing shows a simplified example of a new land use in a Denmark that has been converted to permaculture.

In practice, the planning tools will have to be used on a large scale. For example, networks between cities, the landscape profile with different slopes and horizontal areas, sectoral influences such as soil types and existing natural areas, etc.

Transition, climate and biodiversity:

In this proposal for converting Denmark to permaculture, 64% of the area is used for forests and agroforestry, which produce food, timber and energy.

Biodiversity is increased, clean groundwater is created and carbon is stored. Only 10% is fields, and here food is grown exclusively for humans. Ruminants are kept only for nature conservation. Today, agricultural fields cover 61%.

The proposal is that Denmark must feed 12 million people, so that there is export from zone 3.

The conversion provides a total carbon storage equivalent to 21.3 million tons of CO_2 per year and will result in a strong counteraction to climate change.

	Zones	Yield tons/ha.	Area m^2 pr. person	Ha. in DK total	% of DK area	M Kcal/ ha/year	Bn kcal/year in DK	CO_2-st. new forest/year
Sea	zone 4						170	Large potential
Aquaculture	zone 3						170	
Cities/ infrastructure	zone 0			619,000	14.4%			
Garden	zone 0-1		$83m^2$ x 6m pers.	(50,000)	Under cities	5.19	260	Small storage
Food forests	zone 2		$440m^2$ x 6m pers.	264,000	6.2%	5.19	1370	14 t/ha = **3.7m t**
Nut forests – walnut	zone 3	2.2 t [71]	62 m^2 x 12 m pers	74,000	1.7%	14.39	1060	10 t/ha = **0.7m t**
Nut forests – chestnuts	zone 3	2.8 t [72]	$178m^2$ x 12m pers	214,000	5.0%	4.76	1020	10 t/ha = **2.1m t**
Grain fields	zone 3	5 t	$150m^2$ x 12m pers.	180,000	4.2%	17.5	3150	Small storage
Other fields	zone 3	e.g. 3 t peas	$223m^2$ x 12m pers.	263,000	6.1%	11.0	2890	Small storage
Open habitats/ silvopastures	zone 4			515,000 (Before 387.000)	12.0% (Before 9%)		30	2 t/ha = **0.3m t**
Production forests	zone 4			1,031,000 (Before 558,000)	24.0% (Before 13%)		100	10 t/ha = **4.7m t**
New wetlands	zone 4			100,000	2.3%		20	Stops emissions
Natural forests	zone 5			1,031,000 (Before 75,000)	24.0% (Before 1.7%)			10 t/ha = **9.6m t**
Total				4,295,200	100%		**10.240**	**21.3m ton CO_2/year**

Table 3
Danish example of new land use

Yield measurements have been made at Graham Bell's forest garden in Scotland, which is one of the oldest forest gardens in Europe. The yields have been converted to calories per hectare, and we conclude that 1,650m^2 of food forest must be cultivated in order to have enough calories for one person. This is probably the only thorough yield measurement that has been made on a forest garden in a temperate climate at this time.[67] It is likely that in the future we will see that even higher yields are possible as more experience is gained with growing forest gardens/food forests.

Areas with crops that are rich in starch and oil can produce more calories per unit area than a food forest. This applies, for example, to grains and walnuts. Leafy greens and fruit, which are abundant in a food forest, tend to give lower yields measured in calories. But we cannot be satisfied with fat and starch; we need a varied range of food to be grown so that we can have a healthy and exciting diet.

In our example for Denmark, we conclude that the transition to permaculture farming will mean that we go from 61% of the country being covered by agriculture to only 23% being covered by areas with permaculture farming. With this area distribution, enough food will be produced for one person on 370m^2 of agriculture and horticulture if it is supplemented to a lesser extent with food from the sea, forest and meadow.

In the short term, we could have reduced land use for agriculture even further by continuing to practice conventional agriculture but with the production of only plant-based food. But by switching to permaculture, we solve other problems than just the large land occupation because permaculture stores carbon, supports biodiversity, minimizes nutrient leaching, improves the quality of the soil, minimizes energy input to agriculture, etc. In other words, it offers the completely necessary regeneration that we cannot achieve with conventional farming.

Scenario for new land use

Here we review the different land types, their location in the landscape, and we propose how much land should be used for them in Denmark.

Zone 0 – Densely built urban areas

Densely built urban areas constitute large, and in some cases gigantic, zone 0s, where no significant food is produced. On the other hand, urine is produced, which must be pumped out to zone 3 to complete the nutrient cycle. In Denmark, cities and infrastructure make up 14.4%.

Zone 0–1 – Detached-house neighborhoods

Many Danes live in detached houses and villas with gardens, which are mainly located in the suburbs of towns and cities. The area is relatively densely built up, but each house has access to a garden, some of which can be converted into intensive production of, in particular, annual and perennial vegetables, berries and fruit – in other words, the foods that should be eaten fresh, have a long harvest season and contain a lot of water and are therefore heavy and difficult to transport.

Permaculture points to growing in gardens as part of the food system of the future, since, compared to agriculture, it has the advantage that it avoids some resource-intensive links in the supply chain such as packaging, storage and transport.

In Denmark, there are around 150,000 hectares of garden area. The vision of permaculture is that this should be used to grow food. However, most people would like some of the garden to be used as living space, and some do not want or are unable to grow a garden. We therefore estimate that it is realistic for a third of the garden area to be used for growing food, i.e. 50,000 hectares, which corresponds to 1.2% of Denmark.

Zone 1–2 – Intensive agroforestry and market gardening

Immediately around the towns and cities, we should establish intensive small-scale agriculture with food forests, chicken forests, no-dig market gardens and the like. This will mean that all annual and perennial vegetables, as well as berries and fruit, are grown as close as possible to the consumer in the city. As in private gardens, these are the crops that are most suitable for eating fresh, have a long harvest season and contain a lot of water and are therefore heavy and difficult to transport.

Hazelnuts are also included in these systems because they are good in polyculture with fruit trees and chickens. Ducks are included for slug control, and geese can be included to graze under larger fruit trees for juice and wine.

In this type of area, there is enormous diversity and variation in the systems. This is where the most work is done to optimize the cultivation of perennial crops in several layers and where carbon storage is therefore the greatest.

In Table 3, carbon storage is set at 14 tonnes CO_2e/ha/year in Denmark, i.e. slightly less than the 17 tons measured in southern England (p.38). We have estimated this because Denmark has colder winters and less rainfall, and therefore less plant growth.

In Denmark, zone 1–2 comprises approximately 264,000 ha, which corresponds to 6.2% of the entire country. The area is only intended to supply the current population of six million inhabitants and therefore not produce for export. If Denmark receives large flows of refugees as a result of the climate crisis, this area will have to be increased.

Zone 3

In zone 3, more simplified cultivation systems will be established, producing concentrated food rich in starch, protein and fat, with low water content; this can be transported long distances and stored without damage. We propose that an extra amount of this area be set aside so that there is also enough for export. Cultivation in zone 3 does not require daily care, and much of the work can be done with machines. We divide zone 3 into three different cultivation systems, which are placed in the landscape based on an overall permaculture design that takes into account the landscape profile and sector influences.

Nut forest

Forests of chestnut and walnut trees are placed on areas with a slope, especially the south and west slopes, reducing erosion during heavy rain. Nitrogen-fixing trees are part of the design so that the system can supply itself with nitrogen. In the first 5–10 years, field crops can be grown between the trees. Later, pigs of traditional breeds can be grazed under the trees.

In the land-use proposal, nut forests make up approximately 288,000 ha, which corresponds to 6.7% of the entire country.

Fields for human food

Cultivation of annual field crops such as grain, quinoa, peas, chickpeas, lentils, etc. breaks with the primeval forest as a model and the vision that the entire area in a forest country like Denmark should be reforested, and therefore this should also be limited to being only for human food. It compromises several of the permaculture principles but is appropriate in limited areas, because there has been an enormous improvement in these crops over more than a thousand years. Few of us want a diet completely free of grains. Perhaps in time, when the nut forests have grown and provide high yields of food with similar nutritional content, we can plant some of the fields with trees. But in the first conversion of agriculture, fields are established for human food, and this does not actually take up much of the area.

The fields are placed in areas with a gentle slope and good arable soil. Here, cultivation is carried out in line with permaculture principles, which means that soil tillage is reduced as much as possible, the soil is kept covered all year round by sowing cover crops, and pesticides and chemical fertilizers are omitted.

In the land-use proposal, fields only constitute 443,000 ha, which corresponds to 10% of the country's area.

Aquaculture

Line cultivation of seaweed and mussels in fjords and the sea can be expanded from the current level and have a positive impact on the marine environment.[68] Aquaculture facilities are inspected and harvested relatively few times a year and are therefore located in zone 3.

Zone 4

Zone 4 accommodates extensive systems, where consideration for biodiversity and nature is high, but from which we can harvest in a balanced amount. It is divided into four types:

Silvopastures/open habitats

Existing open habitats are preserved, and new silvopastures are placed – after a sector analysis – on poor and wet soils and along lakes, bogs, streams and the sea. Permanent grass is established here, and scattered trees are planted, so that silvopasture is created. The trees increase carbon storage and provide better well-being for the animals. Extensive grazing, with a low occupancy of cattle, sheep and horses, helps to increase or maintain biodiversity in the grass and will then provide a small yield of meat and milk.

In the land-use proposal, approximately 515,000 ha are prioritized for open habitats, which corresponds to 12%.

Forests for production of timber and biomass

Forests for the production of timber and biomass can be planted on sloping areas, especially the northern slopes, and can be located far away from cities. This area is being increased with a view to our need for more natural materials for building in the future. A small yield of food can also be obtained from the forests, as mushrooms can be produced, fruit and nuts can be harvested from the forest edge, and there will be a yield of meat from hunting.

In the land-use proposal, forests for timber and biomass constitute approximately 1,031,000 ha, which corresponds to 24%.

Wetlands

100,000 hectares of carbon-rich peat soils are being removed from agricultural cultivation, corresponding to 2% of Denmark's area. From lowland soils that are tilled, there is a particularly large loss of carbon and thus an increased climate impact. We must work with nature to restore these wetlands. We propose that some of these areas be used for thatch production.

Sea

From the sea, fish can be caught in balance with the natural environment. Part of the catch is for export, just as it is today.

Zone 5 – Natural forest

Existing forests are preserved. Farthest from densely populated areas and on sloping land, forests of native species are planted, or the areas are allowed to develop into forests on their own. Mushrooms can be collected from these forests over time and a little wild hunting can be done, but all on nature's terms. Today, there is 1.7% natural forest in Denmark. This is expanded in the example to a full 24%.

Based on an overall landscape planning, the untouched forest should be located in zone 5 furthest from the cities. But in addition, space must also be given to smaller pockets of nature in zones 1–4, for example by creating game passages between the intensive small farms, nut forests and fields and by recreating waterholes where it is natural in the landscape. Robust ecosystems will support the cultivation systems, among other things by beneficial animals keeping the population of pests in balance.

Carbon storage through new land use

In the example for Denmark, we reach a full 64% reforestation and the carbon storage through such a conversion of Danish agriculture is enormous. The carbon storage will correspond to a full 21.3 million tonnes of CO_2e per year. This corresponds to 3.5 tonnes per Dane.

In addition, there will be carbon storage in the forests that already exist in Denmark, and carbon storage in the ocean when life in the marine environment increases again due to less pollution from nutrients in agriculture.

At the same time, the scenario will reduce greenhouse gas emissions. Removal of carbon-rich low-lying soils will result in an annual reduction in greenhouse gas emissions of 1.1 million tonnes of CO_2e.[69]

Reducing the number of ruminants to a sixth of the current number will further save the atmosphere from methane, corresponding to approx. 3.9 million tonnes of CO_2e per year, excluding emissions from livestock manure and feed production, and this is where the effect of methane is calculated over a 100-year period. If the impact of methane is calculated instead over a 20-year period, the atmosphere is saved 11.1 million tonnes of CO_2e.[70] Reducing the number of ruminants is thus a climate measure that has a large immediate effect.

The proposal for land use is intended as a discussion paper and shows how far we can go to mitigate climate change by converting to permaculture.

The proposal can be adjusted so that food is produced in Denmark for more people, e.g. 18 million. This can be done, for example, by allocating some of the area for timber and biomass forest and natural forests to food and nut forests instead. With such an adjustment, carbon storage will remain high. This may be relevant because in other countries agricultural land is being lost as a result of climate change, and because the world population is increasing.

How we arrived at the figures is shown in Table 3, p.127.

3 Water

The vision of water in the landscape

In temperate areas, there is usually enough water present in the ecosystems on a year-round basis, and so the task becomes how we manage the water. In warmer and drier climates, where the resource water is in short supply, much more prioritization, planning and often also construction work is needed before permaculture systems can even be established and meet basic human needs. Whilst this falls outside the focus of this book, which is on temperate climates, several of the solutions will also be applicable in such dry areas.

The vision is to get water back into the landscape. Large areas of land, including lakes and bogs, have been drained by laying drainage pipes and digging ditches to make way for agriculture and forestry. But wetlands are the most fertile places in nature, with the most vigorous plant growth, and we can make use of this in our cultivation systems. Water in the landscape is also extremely important for biodiversity.

In permaculture, we want to delay water on its way through the landscape and create reservoirs so that we have water available during drought periods. The water is delayed by collecting rainwater from roofs, establishing ponds and cleaning and recycling wastewater. Finally, the water is stored by making it infiltrate into the soil rather than running off; runoff can cause erosion and, in the worst case, flooding in low-lying areas of the landscape.

Clean drinking water is one of our most indispensable resources, along with heat and food. The quality of groundwater increases when we establish lush plant systems and create healthy topsoil that cleans the water. When we delay water in the landscape, we also increase the infiltration of new clean groundwater. Finally, in permaculture, we reduce the consumption of groundwater by, for example, watering with collected rain-water or treated wastewater.

Water storage in our permaculture systems holds great potential for miti-gating climate change, as good moisture conditions are one of the most important prerequisites for lush plant growth and thus for binding the largest possible amounts of carbon.

Water in the landscape

On the 45-hectare perma-culture farm Krameterhof in Austria, a spring supplies a system of fish ponds con-nected by channels down the south-facing mountainside. Close to the house is a water garden with edible plants.

Soil as a water reservoir

The soil is our primary reservoir. A soil that has a good structure, which is permeable to water, air and plant roots, ensures that rainwater penetrates down rather than running off.

If the water runs off, which can happen during heavy rains and when the soil is already waterlogged, the water quickly ends up in the lowest places in the land-scape and is lost, eventually ending up in the sea via the watercourses. Good soil also has a high content of humus, which can store water like a sponge and thus store it for periods of drought.

Ponds as a water reservoir

The construction of ponds is also a way in which we can increase the storage of water in the landscape.

Groundwater pond

This is a pond where the water level corresponds to the groundwater level. By digging a small test hole before digging the entire pond and following the height of the water level in the hole for a while, we can gain insight into how high the water level in the future pond will be. The test hole should be as deep as we want the pond to be so that it breaks through any stagnant layers.

High-lying pond

If we can build one or more ponds high in the landscape, we can irrigate lower-lying cultivation areas using gravity alone. However, such ponds can rarely be groundwater ponds but need a supply of water and a fully or partially impermeable bottom. If there is heavy clay soil on site, we can simply tamp this down to make a waterproof bottom. Alternatively, clay to seal the pond can be obtained from another place in the local area. Basin foil is also an option, especially for smaller ponds.

Get water for the pond

The most ideal thing is if there is a stream on site from which water can be directed to the pond, and then possibly a whole series of ponds can be made along such a stream. If there is no stream on site, the pond can be supplied with water from hard surfaces such as roofs or rock faces. The runoff from a minor road can also provide a supply of water, but depending on how busy the road is, we must be aware that the water may be contaminated. If we do not have any hard surfaces, we can dig ditches with a slope to the pond from a higher area. However, such ditches will only provide water to the pond during heavy rains and when the ground is already waterlogged.

Distribution of water in the landscape

The best approach to wet and dry areas in the landscape is to work with the landscape, to plant drought-tolerant plants in dry places and plants adapted to moist soil in wet places.

Avoid erosion

If the precipitation penetrates where it falls, the water is initially evenly distributed in the landscape. Therefore, areas with steep slopes in particular must be planted with perennial plants to avoid surface runoff.

It is often enough to establish perennial plants on the ground to solve a problem with water flowing along the surface of the soil. We have ourselves observed that there is no surface runoff water on Reforest Farm, even during heavy cloudbursts on the steepest areas with a 10 degree slope; this is because the ground is covered with grass, trees, shrubs and perennial vegetables that protect the soil, and because this cultivation has created a soil that absorbs the water. When the same land was previously cultivated with annual plants and plowing, there were erosion channels down the hill. In this situation, it has been sufficient to keep the soil covered with living plants, and further measures, such as loosening the soil mechanically with a subsoiler,

A permaculture group from Copenhagen went to Gule Reer twice a week for more than 30 years to grow their fruit and vegetables. At the highest point in the landscape and south of the greenhouse, a rainwater pond has been built. From here, the entire 2 ha area can be irrigated using gravity. Denmark.

South of the forest garden in Holma, a pond has been built to even out the temperatures and reflect the light onto the forest garden. Sweden.

have not been necessary. However other situations may require this, such as when the soil is heavily compacted or in a climate that has intense rainstorms.

If annual plants are grown or there are pastures with grazing animals on steep areas, erosion can be prevented by planting rows of trees that follow the contour lines of the landscape. The trees, with their roots and leaf cover on the ground surface, can keep the soil open so that water infiltrates rather than runs off.

Changing the path of water in the landscape[1]

The 'keyline system' developed by the Australian P.A. Yeomans uses a 'keyline plow' that makes underground channels in the soil to increase water infiltration. The channels are made along the contour lines, but with a slight deviation so that they have a slight slope. This prevents the water from taking the shortest path down the hill to the lowest areas, where there is already plenty of water, and we thus avoid water erosion and achieve a more even distribution of water in the landscape. In some situations, the channels can be used to direct the water underground to where you want it.

On a large landscape scale, it is sometimes possible to direct water from a higher-lying depression in the landscape with excess water to a lower-lying ridge where it is dry.

Here you can see erosion channels between harvested rows of corn in the winter of 2013 in the area where the Reforest Farm now grows a food forest. The soil now absorbs all the water during heavy rains.

Cultivation of water-logged soils

Areas with high groundwater levels are particularly challenging in terms of cultivation. The vast majority of plant species cannot tolerate water-saturated soil because they need oxygen for their roots, and if we plant them in a place with a high groundwater level, they will develop a superficial root system, which then makes them vulnerable when the groundwater level drops during a drought.

If we only have the opportunity to cultivate such a place, we can choose to drain the soil, but this is not a very good solution because we then lose the water that the plants need during a drought. A better solution is to make raised beds for individual trees and shrubs or for our vegetables. This gives the plant roots more soil to grow in above the groundwater table. Soil for the raised beds can be obtained, for example, by digging a pond on site.[2]

Turn lowlands into nature

The climate crisis has, with good reason, created a lot of focus on the carbon-rich lowlands, i.e. meadows and bogs that have been drained and used for

At Ridgedale, a 'keyline plow' was used on the 8 ha area to create a good distribution of water in the landscape. Rows of fruit and nut trees were then planted, between which cows and chickens are now grazing. Sweden.

The pond is built high in the landscape with a membrane and is important for biodiversity. Sunlight hitting the surface will be reflected onto the solar cells and increase their production. The treated gray wastewater from their wastewater greenhouse, together with roof water from the house, flows into the pond. The pond overflows in the event of heavy rain. The Permaculture Roots in Skovvirke. Denmark.

agricultural land. Enormous amounts of CO_2 and nitrous oxide are released when such areas are drained, and oxygen is added to the large amounts of carbon they store.

A meadow or bog that is under water releases the greenhouse gas methane, but in relatively small amounts; it therefore has a much lower climate impact than if it were drained.

By stopping the drainage and restoring the wetlands, a very large reduction in greenhouse gas emissions of up to 40 tons CO_2e/ha/year is achieved,[3] and at the same time we return the area to nature.

Water in the house and garden

We can save groundwater by reusing water in the household and by replacing groundwater with collected rainwater.

Toilet flushing accounts for a large part of our water consumption, approximately 30 liters/person/day. If we have a composting toilet, we avoid this water consumption and at the same time separate urine and feces from the rest of the household water.

Wastewater containing urine and feces is called black water, and wastewater without urine and feces is called gray water. Gray water contains only a few nutrients and far fewer bacteria than black wastewater, so it is easier to clean.

Cleaning and using gray water

By taking care of our wastewater ourselves rather than directing it to an energy-consuming public treatment plant, we have the opportunity to make use of this water resource ourselves. This is an example of delaying the water's path through the landscape.

Unlike rainwater collection, the flow of wastewater from a household is constant throughout the year, so if you treat the wastewater before use, it can be a bit like having a spring on your property.

Below are some examples of the treatment and use of gray water. When designing wastewater solutions, it is important to investigate what is permitted in terms of legislation in general and in your location.

Separate piping to a greenhouse

To make the best use of gray water for irrigation, we want it to be spread out as widely as possible. This will also allow as many microorganisms as possible to clean it. Instead of collecting the pipes in one large pipe, it may be an idea to keep the pipes from the bathroom sink, kitchen sink, washing machine and shower separate and route them under various shrubs and trees around the house and then cover the pipes and

The greenhouse cleaning system receives gray water from the house through a separate pipe from each source. The pipes end in a 'box trough', which is a box with a lid and no bottom. The water from the kitchen sink is pre-cleaned by a basket of straw inside the box trough (the straw has to be changed from time to time). From the box trough the water flows vertically through a layer of pebbles to the bottom of the system that is lined with a plastic membrane. From here some of the cleaned water can be used by the plants in the greenhouse, and the rest leaves the greenhouse ready to enter the garden pond. The system uses no electricity, saves water for irrigation, and the plants benefit from the nutrients. Reforest Farm in Skovvirke. Denmark.

their outlet with a layer of straw to keep them frost-free in winter. If we want the water to be even more spread out, it is possible to put a flow splitter on the pipes so that they branch out and have even more outlets.[4]

This simple way to make use of wastewater saves investment in a septic tank, is cheap to install and robust in operation, but it has a low utilization rate of the water for irrigation.

An optimization of such a system can be to let the pipes drain into a greenhouse. In the greenhouse, there is a particularly high need for irrigation, as there is no rainfall.

Since there is a higher temperature in the greenhouse than outside, the greenhouse can help ensure that the pipes do not freeze, and the higher temperature also improves the working conditions for the bacteria that are supposed to purify the water.[5]

Reedbed system

Another option is to clean the wastewater before it is used for other purposes. A vertical-flow reedbed consists of a bed of small pebbles planted with common reed. Microorganisms living on the pebbles and roots of the plants clean the water. It can be used for black and gray water. It can be smaller if it is only fed gray water, and with a composting toilet that keeps urine and feces out of the wastewater, you can make better use of the nutrients from the household.

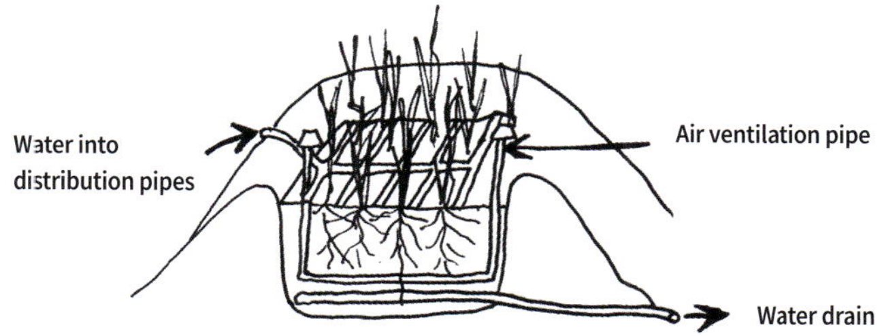

The vertical-flow reedbed cleans the water with the help of microorganisms on the surface of the pebbles and plant roots. The water that comes out of the drain can be used for irrigation or for a pond. Denmark.

RENE KILIAN

Before the water enters the vertical-flow reedbed it passes through a septic tank where solids settle and oil and grease float to the top. From here the water is pumped into the reedbed usually with an electric pump; by pumping in the water it ensures the water is distributed evenly over the surface of the system. If there is a 2-meter drop before the reedbed, the water can alternatively be pumped with a siphon pump that does not use electricity.

The vertical-flow reedbed cleans the water to bathing-water quality and ensures that it meets the WHO's requirements for the reuse of purified water for watering field crops.[6] The water should therefore be safe to use for garden irrigation or to add to a pond.

Unfortunately, the authorities are often very restrictive about the use of purified water due to the fear that people will come into contact with disease-causing bacteria.

Rainwater collection

In many places, it is an obvious option to collect rainwater from the roofs of buildings, and it can be a good idea to do so with a much larger capacity than a single rainwater barrel so that we have ample amounts of free water for watering, flushing toilets and running washing machines, etc. Using rainwater for laundry means that the consumption of soap can be reduced because rainwater is soft water.

Excess rainwater can be directed to an irrigation bed with, for example, reeds or bulrush, which can be harvested and used for mulch in no-dig cultivation.

Calculating needs and collecting rainwater [7]

In order to avoid using groundwater, we must calculate our needs and how much we can collect.

Need: As a rule of thumb, it is best to irrigate once a week during a drought and to irrigate 25 liters per square meter per irrigation. If you want to be able to handle a three-week drought period in a $100m^2$ vegetable garden, you will need

$$100m^2 \text{ x } 25 \text{ liters x } 3 \text{ weeks} = 7{,}500 \text{ liters or } 7.5m^3.$$

Collection: The amount of precipitation that falls varies by location, even within individual countries. You must look at a precipitation map to find out the average annual precipitation where you live. If, for example, 600mm falls per year, this corresponds to 600 liters of water falling on each square meter of the house's floor area.

Some of it evaporates again before it reaches the downspout, so you must multiply by a factor of 0.7 if you have a pitched roof, and 0.5 if there is a flat roof. In addition, there is a minor loss if there is a filter in the downspout to keep coarse dirt out of the tank, so you must multiply by a factor of 0.8 or 0.9, depending on which filter it is.

If the house has a floor area of $90m^2$, the roof has a slope, and the water runs through a filter, the calculation looks like this:

$$90m^2 \text{ x } 600 \text{ liters x } 0.7 \text{ x } 0.9 = 34{,}020 \text{ liters per year or } 34m^3.$$

The fire pond at Sieben Linden, located 200km from the sea, is designed as a biotope with lots of life and also functions as a well-visited swimming pool. Germany.

Aquaculture

There are some basic ecological conditions that mean that food production in water can be far more productive than on land.

- Lake shores have a pronounced edge effect and thus great fertility. Plants that grow in shallow water have both water, soil and air available and never suffer from drought stress. Foliage from trees like alder is also an important source of fertilizer for plants in the water, just as insects that fall from the trees are an important food source for fish.
- The nutrients in the water are in dissolved form and are easily accessible to the plants.
- The water evens out the temperature, as in a coastal climate.
- Fish are cold-blooded and do not use energy to keep warm.
- Fish do not need to use energy to support their own body weight as they are supported by the water.
- Fish move up and down in the water to niches with different food resources, which is an example of the permaculture principle of stacking.

The pond is artificially constructed with a pond liner and functions as a habitat for insects and amphibians. Watermint, arrowhead and watercress are grown here. The warm microclimate at the pond is used for grapes and figs.

Watercress is planted in pots with a mixture of sand, soil and compost. Pebbles are placed at the top so that the pond does not become muddy.

The water garden

Creating a garden pond provides both a beautiful element in the garden and is one of the best things that can be done to support wildlife, including salamanders, frogs, toads, dragonflies and birds.

Aquatic plants

At the same time, garden ponds can be used for vegetable production and can be very productive. Some of the best edible aquatic plants to grow are arrowhead *Sagittaria* spp., watercress *Nasturtium officinale* and bulrush *Typha latifolia*.

Arrowhead is grown for its root tubers, which are rich in starch. The tubers are 1.5–3cm in diameter and are prepared by boiling them for 5–10 minutes.

Watercress is used as a salad, is rich in vitamins, calcium and iron and can be harvested not only in summer but also in winter in a garden pond, as long as it does not freeze.

Bulrush is also a vegetable. Here the edible shoots are harvested in spring, and later on, the young flower heads. The starchy roots of the bulrush can also be boiled and eaten. It is a plant that spreads rapidly, and it is best to grow it in a container or a bed by itself.

Water fern *Azolla* spp. and duckweed *Lemna* spp. are floating aquatic plants and are some of the most productive plant species that exist. Both have a high protein content of around 25–35% of their dry weight, and water fern is also a nitrogen fixer. They can be harvested by skimming them from the surface of the water and can be used as feed for chickens and other animals, but the plants must be dried first, otherwise the animals fill themselves up with water. Water ferns are not winter-hardy in cold climates and must be kept indoors over the winter. In some areas of the world, *Azolla* is a very problematic invasive species and must not be cultivated. The sale of *Azolla* is banned in the UK.

> We must ensure that sheep, cows, deer and other grazing animals cannot get near the garden pond where we grow watercress and other vegetables that are eaten raw. The excrement from these animals can contain the eggs of liver flukes, which are very harmful to human health.

At the bottom of the pond, swan mussels are placed to clean the water. Reforest Farm, Denmark.

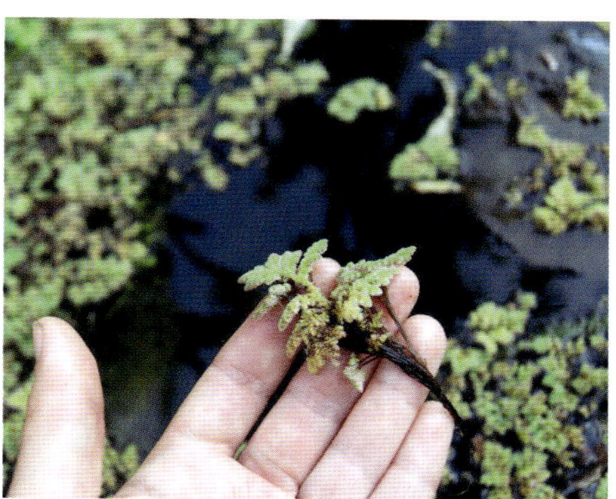

Water fern is nitrogen-fixing and very productive, and can be harvested and used as fertilizer.

Aquaculture with carp

While it is relatively simple to make a garden pond and produce your own water vegetables, aquaculture with fish is more complex. On the other hand, it also offers the opportunity to make many beneficial relationships and achieve high production on small areas. By producing fish, we also create an alternative to eating fish from the sea, where many fish stocks are currently in danger due to over-fishing.

Carp are often the best choice of fish for aquaculture in temperate climates because they are low in the food chain. For example, common carp eat both small animals and algae in the water. Carp can be produced in stagnant, warm and relatively oxygen-poor water, which means that carp production does not require that we have a watercourse on our permaculture site.

Trout and salmon, which are common fish in modern fish farming, are predatory fish that depend on high-protein feed. A stream with cold, oxygen-rich water is needed to farm these species.

If we want to create a productive system with carp, we need three ponds. Carp take three years to grow to a size ready to eat, and the fish are therefore divided by age (one, two and three years old) as otherwise there will quickly be too many small fish that take food from the larger ones, which then do not grow very much.

The combination of fish and aquatic plant production is ideal because the fish's excrement provides nutrition for the aquatic plants, and the aquatic plants help increase the oxygen content in the water for the benefit of the fish. The pond should then contain both shallow areas with aquatic plants and areas that are so deep that the bottom does not freeze, so the fish can overwinter in the pond.

An intensified version of fish aquaculture is to have different ponds, between which the water is circulated by a pump. The ponds may contain a clarification pond, where dissolved material can settle, then a pond with daphnia, water fleas and other small animals, which flows into one or more ponds with fish (via the overflow), and then an aquatic plant pond that utilizes the nutrients from the fish excrement, before the water is pumped back to the start.[8]

Markus Vallin in Sweden has created several ponds for the production of carp. The fish are sorted by age so that they can grow large enough.

The systems can also include fountains or water stairs where the water is oxygenated. The disadvantage of such systems is that they are dependent on electricity for the pump. It may be worth trying to base the water circulation, in whole or in part, on a solar pump that functions when the sun is shining.

Aquaculture and methane

An unfortunate effect of aquaculture systems is that they contribute to climate change by emitting the powerful greenhouse gas methane. Methane is produced when plant material and animal manure are placed under water so that the reaction takes place under oxygen-free conditions.[9, 10]

One approach to reducing methane emissions from an aquaculture system is to reduce the amount of dead organic matter in the water to prevent it from rotting. The aquatic vegetables are harvested, and wilted leaves and other

In Knud Henrik Koefoed's garden, ponds with the plant water fern have been created among berries, fruit, vegetables, chickens and ducks. Denmark.

Under the fruit trees in Knud Henrik Koefoed's courtyard, flowers and herbs grow in beds between the fish ponds with carp. Denmark.

CATHRINE DOLLERIS

By using seagrass for our cultivation, we are taking in the excess nutrients that have ended up in the sea. Cathrine Dolleris, Denmark.

dead plant parts can be removed from the water. There is also some evidence that planting strong plants with hollow stems, e.g. bulrush, can reduce methane emissions because they carry oxygen down into the bottom sediment.[11] Similarly, artificial aeration of the water is thought to reduce methane emissions.[12]

Coastal ecosystems

Coastal ecosystems, such as tidal marshes, kelp forests, seagrass meadows and, in tropical areas, mangrove forests, are important habitats and very productive ecosystems with a primary production per square meter that, in some cases, can be on a par with forests on land.[13, 14]

Part of the carbon is stored in the seabed as 'blue carbon', and it is released slowly due to the oxygen-free conditions. A global study estimates that if everything is done to protect endangered seagrasses, salt marshes and mangroves and to restore lost populations, it could provide a carbon sink and prevent emissions equivalent to approximately 3% of annual global emissions from fossil fuels.[15]

Seagrass is being reestablished

Eelgrass (*Zostéra marina*) is the most widespread seagrass in the northern hemisphere. Eelgrass grows to a depth of 11–17 meters if the water is clear. In Denmark, at least two-thirds of the area of eelgrass that grew along the coasts and in fjords has disappeared since 1900, partly because nutrient leaching has created algae growth that makes the water less clear.[16]

Where nutrient pollution is less intense, it is possible to plant eelgrass and thereby recreate vigorous-growing beds. Today, groups of volunteers around Denmark are involved in replanting the lost seagrass meadows.

In addition to the fact that replanting creates habitats for marine life and increases carbon storage in the ocean, eelgrass is a valuable resource for us humans. Eelgrass can be collected when it washes up on the beach and used for many things, including for mulch and insulation material in houses. In this way, we remove excess nutrients from the marine environment and store carbon.

Aquaculture with mussels and seaweed

Unlike wild fish stocks in the ocean, which are often overexploited by fishing, it is possible on a global scale to greatly expand food production from the sea through the production of mussels and seaweed. This will not only contribute to a healthier diet, but we can also do it in a way that helps create a healthier marine environment. Mussels and seaweed can be grown on lines that are hung out in the water. Such production does not require any feed or fertilizer, but does absorb nutrients from the sea.

HAVHØST/LARS NORMANN HESTBÆK

Long-line mussel farms produce blue mussels, which is a healthy food, while also removing excess fertilizer from the sea and storing carbon in the mussel shells. Natural resources are regenerated and climate change is mitigated.

Cooked in a pan, the brown seaweed turns bright green. Like so many other vegetables, seaweed can be prepared in all sorts of ways: raw, boiled, fried and baked.

By growing mussels on lines, we avoid scraping the seabed as we do when catching wild mussels, and at the same time we get a better product in the form of meatier mussels. Mussel farms are located in fjords, sea lochs and sheltered bays, where the weather is not as severe as on the open sea. A long-line mussel farm typically measures 250 × 700 meters, and a farmer usually has at least two such farms, as a production cycle takes two years.[17]

Just like when harvesting eelgrass, harvesting mussels and seaweed removes some of the nutrients that are otherwise too abundant in the sea.

Mussel shells contain carbon, and if the shells are used for building material, for example, we have created a carbon store, and climate change is mitigated.

The organization Havhøst (seaharvest) works in Denmark to spread the cultivation of edible crops in the sea, both in the form of larger farms and on a smaller scale for associations and individuals who grow mussels and seaweed for their own consumption in so-called maritime utility gardens, i.e. vegetable gardens in the water.

HAVHØST

Seaweed has been used as food and animal feed since the Viking Age. In Japan, seaweed makes up 10% of the diet. Seaweed contains large amounts of vitamins A, B and C, iodine, minerals and omega 3 fatty acids. Seaweed is harvested here by the organization Havhøst.

Seaweed can be harvested on rocky beaches. It is tastiest in early summer, as small algae begin to grow on it later.

4

Energy

The vision of regenerating energy

The permaculture vision is to create a local energy system that regenerates natural sources of energy, helps to mitigate climate change and enables us to live comfortably. Energy consumption must also be reduced through increased energy efficiency in our homes, transport and food system. The wealthiest part of the world's population today has a large overconsumption of energy, and we must reduce this. Energy that is not used does not create climate change, does not occupy land and does not consume raw materials.

Regenerating the resource energy

Regenerating the natural sources of energy means first and foremost creating more lush plant systems. For example, we can increase the amount of wood in our forests by harvesting less timber than the forest grows each year.

In addition to firewood, plant systems can also produce the raw materials for the production of ethanol (alcohol), vegetable oil or biogas, all of which are high-quality fuels that can be used for many purposes. But we need to be aware that the way in which such fuels are produced today and the amount of them used leads to a degradation of the natural sources of energy. For example, when palm oil is used for biodiesel, which indirectly causes the clearing of forests, bogs and peatlands that have large carbon stores, it means that biodiesel made from palm oil has a greater climate impact than fossil diesel.[1]

Building a pond high in the landscape is also regenerating the energy resource by virtue of the potential energy of water. This energy can do work for us if we direct the water to a turbine further down the hill.

Increasing the production and storage of food is also an example of regenerating energy in the sense that food is energy for people, but we have discussed this topic in the chapters on agriculture and cultivation systems.

Technologies such as wind turbines and solar photovoltaic panels do not directly build up the natural sources of energy. It takes large amounts of raw materials to produce them, and the solar panels can take up a lot of land if they are set up as solar farms out in the countryside. But indirectly, wind turbines and solar panels help build up the natural resource of energy because they can 'harvest' energy without us having to cut down forests.

An energy system that stores carbon

Burning biomass for energy purposes is in a way CO_2-neutral, because the amount emitted when it burns equates to the amount absorbed by the plant when it was growing. But if we regenerate the resource energy in the plant systems, it also means that we store carbon and thus mitigate climate change.

In the Karise Permatopia ecovillage, the goal is to be self-sufficient in energy with a closed loop, so that the use of waste electricity is optimized. The homes and the farmhouse that is used communally are heated with a ground source heat pump; the windmill produces electricity for the heat pump, the farmhouse and the electric-car charging stations; the excess heat from the heat pump contributes to heating the hot water. Denmark.

PETER ARENTOFT JAKOBSEN

Emergy

We can gain a deeper understanding of what energy is through the concept of 'emergy', which was developed by energy researcher Howard Odum. The 'm' comes from the word 'embodied', and we can think of emergy as 'energy memory'. Emergy refers to the energy that has been used directly or indirectly to make a product or to make a service available. Things with high emergy are worth more and can do more than things with low emergy.

Odum was one of the inspirations for Bill Mollison and David Holmgren in their development of the permaculture concept.[2]

The emergy table on the right is a slightly shortened version of a table from Howard and Elisabeth Odum's book *A Prosperous Way Down*. The table is based on solar energy, as most energy stores and energy flows on earth are created by solar energy; tidal energy, geothermal energy and nuclear power are significant exceptions.

Wind is created by the sun; the atmosphere in some areas is heated more than in others, thereby creating high pressure and low pressure and thus wind. From the table we can see that it takes about 1,500 calories of sun to create 1 calorie of wind.

Trees and other plants also grow using solar energy, and here it takes about 4,400 calories of sun to create 1 calorie of biomass.

Fossil fuels are plants and small animals that have been under high pressure for a very long time. Only a small part of the plant production has become fossil fuels, and therefore the ratio here is 50,000 calories of sun to 1 calorie of fossil fuel.

A material or service with high emergy is of a higher quality than one with a low one. It is more valuable and has more applications.

Electrical power, for example, has a higher emergy than coal. At a coal-fired power plant, about 4 calories of coal are used to produce 1 calorie of electricity. Electric power, on the other hand, can be used with a higher energy efficiency than coal, has more applications such as lighting, computers, etc. and is easier to transport than coal. For example, electric trains are more energy efficient than trains powered by coal.

But what can we use this knowledge about emergy/energy qualities for in practice? We can derive a general guideline, which is that we should avoid using high-quality energy for low-quality purposes, as it is simply wasteful. This is how nature's systems work, and they have optimized their function over eons.

Today, it may not seem as if electricity from solar cells and wind turbines is so precious, even though it has a high energy quality, because it is currently competing with electricity produced by burning fossil fuels. But it is still a big challenge and very resource intensive, to get solar and wind energy to do the same as fossil fuels, for example to be available all the time even when the wind is not blowing and the sun is not shining, and to power airplanes, cargo ships, trucks and tractors.[3] So in that way, it is a precious resource.

While we must avoid using high-quality energy for low-quality purposes, we must also choose the energy sources that have the least climate impact. For example, if the electricity comes from renewable energy sources, it is better to use it than to use oil, even if the oil has a lower energy quality.

Everyday examples of applied emergy

Heating a house with electric radiators is using high-quality energy to get heat (low-quality energy). We can easily keep a home warm with energy forms of a much lower emergy; by getting solar energy in through south-facing windows or by burning wood when the sun is not shining.

If we want to use electricity for home heating, we can get more out of the precious electricity by, for example, using a heating system based on a heat pump, which will give us approximately 3–4 times as much heat per kWh of electricity as the electric radiator. Here, the electricity is used in a more advanced way than in the electric radiator, namely to pump heat out of the ground or the outside air and deliver it to the home. We can only do this with high-quality energy such as electricity.

Cooking is often done with electric hobs and thus consumes high-quality energy. We can cook just as conveniently over natural gas or biogas. But what is most climate-friendly depends on how the electricity is produced.

Wood-burning stoves are also an option, although they are less convenient than gas and electricity. Firewood is a lower-quality product, but it is easy to produce locally.

Energy quality/emergy[4]

Calories of solar energy previously transformed directly and indirectly to produce a calorie of energy of the type indicated.

	Solar emcalories per calorie
Solar energy	1
Wind energy	1,500
Organic material, wood, soil	4,400
Energy from large river	40,000
Fossil fuels	50,000
Food	100,000
Electric power	170,000
Protein food	1,000,000
Human service	100,000,000
Information	100,000,000,000

Historically and in the present

Historically, traditional biomass has been the most important energy source, and our society functioned with very small amounts of high-quality energy and relatively simple energy technology until the 18th century. Here, the discovery of fossil fuels gave us access to large amounts of high-quality energy and, with them, a more advanced society in many ways.

Today, work is underway to transition to renewable energy technologies, which, just as before fossil fuels, are based primarily on sun and wind. But in order to be able to convert low-quality energy such as sun and wind into high-quality energy in the form of electricity, far more advanced energy technology is needed than what we had previously.

Decentralized low-energy society

The way we obtain our energy today is one of the main causes of climate change. The current energy mix still largely consists of climate and environmentally harmful energy sources such as coal, oil and gas, which together accounted for 85.3% of global energy consumption in 2023. The same year, solar power accounted for 1.0%, wind for 1.4%, hydro for 2.5% and traditional biomass for 11.1%.[5] Although solar and wind power are increasing rapidly, there is still a long way to go to achieve a climate-friendly energy system, and the energy moves around in a global energy market, so it is the same system that we get our energy from.

Local energy harvest

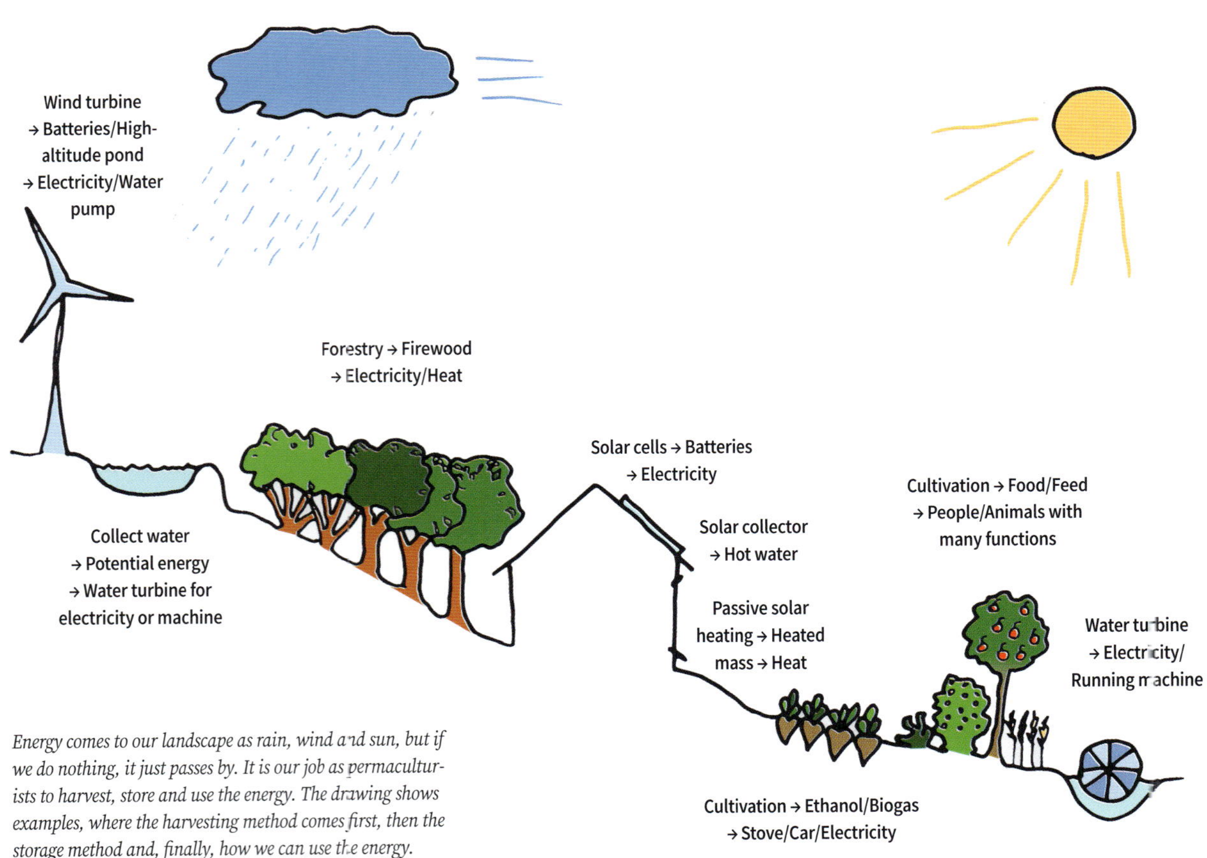

Energy comes to our landscape as rain, wind and sun, but if we do nothing, it just passes by. It is our job as permaculturists to harvest, store and use the energy. The drawing shows examples, where the harvesting method comes first, then the storage method and, finally, how we can use the energy.

Limit energy consumption

The political system has failed, and we have started the transition to renewable energy too late. There is simply no more time left in relation to the climate crisis, and the only responsible thing is therefore to cut energy consumption here and now. Reducing the amount of fossil fuels that are burned has an impact, and we can all contribute to this by reducing our current consumption.

Unfortunately, we in the general population lack awareness of the extent of the crisis we are in and the necessity of changing lifestyles to reduce energy consumption. As permaculturists, we can take the lead by showing in various ways that it is possible to maintain a high quality of life with much lower energy consumption.

Some of the most obvious ways to cut back are to insulate our homes, take trains instead of cars and planes and use a clothesline instead of a dryer. We may also have the option of working from home or living close to our workplace so that we avoid commuting. We also have an influence on energy consumption in sectors such as agriculture and industry through the goods we buy and do not buy. If we use less energy, the renewable energy plants that have already been built will be able to cover a larger share of our consumption.

Transition to renewable energy

Considering the climate crisis, there is no doubt that the large energy companies and their engineers must deliver what they can to build an energy system based on renewable energy in the form of new wind turbines and solar parks. But the rest of us must also get involved; the matter requires a strong commitment from society as a whole. If we ourselves help produce what we need at the household level, locally or bioregionally, we can help democratically manage development and completely exclude fossil fuels. We will also experience the negative consequences of our way of using energy, for example how much forest is cut down, as well as aesthetics and noise in connection with the installation of wind turbines and solar cells. Finally, we also get the benefits of the local jobs associated with renewable energy plants.

As permaculturists, we can get involved in the production of energy either by joining a local community energy project, like a wind cooperative, or take an even more far-reaching step by going off-grid. Being off-grid means that you refrain from connecting to the electricity grid and possibly also other networks for gas, district heating, water and wastewater, and instead create local solutions in line with the principle of small intensive scale.

On the grid or off-grid?

There are good reasons to establish a coherent electricity grid, because resources of solar, wind and hydropower in different parts of the world can complement each other. At one point, there may be a surplus of wind energy in Denmark, which can be sent to Germany, where there is a shortage of renewable energy, or a surplus of hydropower in Norway can cover a deficit of energy in Denmark. This reduces the need for energy storage or for obtaining energy by burning something. If you are off-grid, you are more dependent on storing the electricity, which is usually done with batteries, which costs money and has an associated environmental impact.

But there are also good reasons why more and more people are choosing to go

EMMA H. NIELSEN

The off-grid house at Brenderup Folk High School in Denmark is a house that is simply placed inside a greenhouse so that the entire microclimate around the house is warmer and the heat loss is therefore lower.

off-grid, especially in the countryside and in smaller towns, where there is room for their own harvesting of solar and wind energy.

Harvest and utilize energy locally

Unlike fossil fuels, renewable energy is distributed over the globe in a thin layer. In relation to biomass and hot water from solar collectors or combined heat and power plants, it is an advantage to utilize them in the same local area as they are produced, in order to create the most efficient energy system with short piping for hot water and short transport distances for biomass.

Avoid the rebound effect

People who are off-grid become good at adjusting their energy consumption according to when the sun shines and the wind blows, and they are usually very aware of limiting their total energy consumption, because energy is a limited resource. This is of great importance, as research has shown that energy savings, such as additional insulation and the like, usually lead to increased consumption, for example by turning up the heat or living in a larger home. This is called the rebound effect and is the reason why the total energy consumption in society does not decrease, even though we improve energy efficiency.

Resilience

Greater resilience is also a key reason to go off-grid. We have designed our lifestyles to be extremely dependent on uninterrupted supply from the electricity and water grid. The more of us who are off-grid in our local communities, the better we will be able to cope together if a prolonged supply disruption occurs.

Financial resilience

Price increases are another impact that we become resilient to when we are off-grid. The supply grid is affected by events elsewhere in the world that can cause energy prices to explode. When we are off-grid, we know the costs, which are primarily for installation.

Wood for energy purposes

New coppice woodlands mitigate climate change

An example of a plant system that can be established for energy purposes is the coppice woodland. In this type of woodland, the same trees are cut down to the stump or stool at intervals of 8–15 years and then shoot again with multiple stems. In addition to the fact that a coppice woodland can produce wood for fuel almost infinitely, a newly planted coppice woodland will also store carbon in the ground in humus and root mass and on the forest floor in the form of dead leaves. Wood can be harvested from the same trees in a coppice woodland for millennia, and during all that time, carbon stored in the forest floor is prevented from escaping into the atmosphere as CO_2. The coppice woodland is also a very valuable habitat for wildlife, and in addition to firewood, the coppice can supply us with poles for construction, fencing and craft.

We must choose tree species for the coppice woodland that are good at regrowing after being cut down, and also preferably some that are native and good at supporting biodiversity. In northern Europe, this could be hazel, ash, field maple, common alder, lime, poplar, willow and sweet chestnut (not native). Birch can also be used, but mostly as a self-sowing species. Oak can work too. When establishing a coppice woodland, the trees are typically planted about 2 meters apart (depending on species).

Particularly high yields can be achieved by establishing a short rotation coppice of densely planted, high-yielding varieties of poplar or willow. Short rotation coppice is usually harvested every 3–4 years. Unlike traditional coppice woodlands, the area is replanted after 20–30 years due to declining yields, and this makes the system's value as a carbon store somewhat less. In cases where large monocultures of willow or poplar clones are planted, they are not particularly beneficial for biodiversity either.

Bjarne Wickstrøm in Denmark is self-sufficient in fuel from high-yielding varieties of willow. The willow is harvested at short intervals of 3–4 years, and a large yield is achieved. Short rotation coppice is often replanted after 20–30 years due to declining yields.

Wood – a limited resource

If the growth in sustainably managed forests cannot keep up with the burning, the burning contributes to worsening climate change. An assessment indicates that globally, 10GJ of biomass is available per world citizen per year if the resources are to be managed sustainably.[6] This corresponds to approximately 500kg of firewood, both for heating and electricity.

Black carbon – climate impact from burning

In addition to the greenhouse gas CO_2, there is also a climate impact from particle pollution, also called 'black carbon', when we burn something, especially in the case of unclean burns. The particles absorb heat in the atmosphere, and when particles land on snow and ice-covered areas, they will darken the color of the snow so that it absorbs more solar energy than otherwise, resulting in melting, which in turn exposes the ground, which can absorb even more heat, i.e. a self-reinforcing effect. Black carbon's impact on ice-covered areas is of particular importance when the burning takes place close to the ice-covered areas. Since black carbon only stays in the atmosphere for a few days to a few weeks, reducing emissions will immediately result in a reduction in global warming.

The Hunter Stove is a biochar stove with a chimney, and could possibly be adapted for indoor use.[7]

Energy technologies

As mentioned under the permaculture principle of harvesting energy, it is our task as permaculture designers, based on the energy resources that flow through our place, to harvest and possibly create stores for the energy, where it is in the form that is most useful to us.

In order to harvest the energy, transform it and store it, we need different technical solutions, and these have different advantages and disadvantages and are suitable for different situations.

In general, we can say that there is a great need for more development of small-scale energy technologies and for more practical experience in using them in permaculture contexts. We have selected six specific and different solutions which are highlighted below.

Biochar stoves and ovens

Biochar stoves and ovens may prove to be an important element in future permaculture systems because they rely on technology that can be used to directly mitigate climate change.

Wood or other biomass is burned with reduced air supply, which is called pyrolysis, and the heat from the combustion can be used for various purposes. During pyrolysis, gas is released from the wood, while up to half of the carbon in the wood is retained in the form of biochar (charcoal), which is left after combustion. The biochar will constitute a stable carbon store that can last for 100–1000 years and during that time counteract climate change. The large range in durability depends on both what the biochar is made of and the temperature during pyrolysis. Biochar becomes most durable when the pyrolysis takes place at temperatures above 500°C.[8]

Small biochar stoves, which we can put a pot on top of, are relatively widespread in developing countries. These are often of the top-lit updraft type (TLUD). The TLUD consists of a container that is open at the top and has small air holes at the bottom.

The container is filled with wood or other dry biomass, and a fire is lit at the top. The smoke rises to the top of the container where it meets preheated air, which rises in a jacket on the outside of the container. Where the gas meets the hot air, it is burned off. The TLUD has clean combustion and low emissions of particles and black carbon.[9] The biochar stoves that have been developed so far are intended for outdoor cooking and are therefore not very useful in

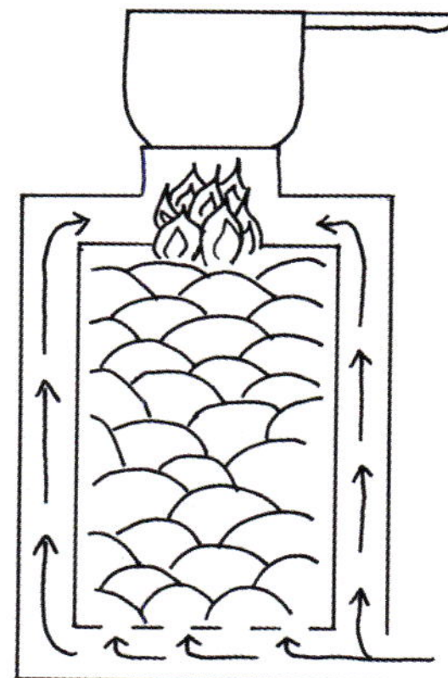

The TLUD stove has a primary air intake through a grate under the burning material and a secondary air intake by the flames just below the hob. This ensures that the flue gases are burned.

temperate climates. If we are to seriously integrate biochar stoves and ovens into permaculture in temperate climates, the technology must be developed for indoor use.

At Edible Acres, a forest farm in New York State in the USA, they use a regular wood-burning stove with a large combustion chamber for both biochar production and for heating the home at the same time.[10] A glowing fire is made in the wood-burning stove. A metal container filled with pieces of wood and with a loose lid is placed on top of the glowing fire. The heat from the embers causes the wood to degas, and gas rises along the edge of the lid. This gas burns when it comes into contact with oxygen inside the firebox. It is important that the lid is loosely fitted, otherwise gas pressure can build up and there is a risk of explosion.

A step further is to develop completely new types of stoves that are intended for indoor biochar production. Edward Revill, a permaculturist in Wales, has developed the 'Biochar Rocket Stove'. The stove is built around an angular tube. The horizontal part of the tube functions as a firing tube and the vertical as a heat riser, which creates a draft in the stove. Outside the heat riser, a separate chamber has been built, which is filled with biomass.

When the stove is fired in the angular tube, the biomass in the chamber is heated, and it begins to degas. The gas penetrates the smoke riser through some cracks and is burned off. After firing, the outer chamber is emptied of charcoal.[11]

Flue gas scrubbers and filters

As previously mentioned, particle pollution from burning wood is both harmful to health and contributes to climate change. Therefore, it is first and foremost important to fire correctly using dry wood in a good stove. It is also a good idea to clean the smoke using a flue gas scrubber or filter. The flue gas scrubber is being developed by the Danish inventor

The rocket stove mass heater in Birkegården's gardens has no chimney because the flue gas is led under the floor to the flue gas scrubber, which is made of a blue barrel with LECA balls in it. Here, the flue gas with the nutrient salts in it condenses, and the barrel is filled with nutrient water over the winter.

BOTH: CATHRINE DOLLERIS

and straw-bale house builder, Steen Møller. The purpose is to clean the smoke from wood-burning stoves of particles, to extract the nutrients that otherwise disappear through the chimney as fly ash and to increase the utilization rate of the wood that is burned. The system first consists of a smoke channel or a hollow wall, where the smoke from the stove is cooled down so that the water vapor in the flue gases condenses. When the water vapor condenses, energy is released, and the energy utilization of the wood is increased. At the end of the smoke channel is a fan that provides the necessary draft.

The flue gases then enter the flue gas scrubber itself, which consists of a plastic container with a substrate in it that the smoke must pass through. Various substrates have been tested, including LECA (lightweight expanded clay aggregate) balls, charcoal and sphagnum. The substrate is kept wet by pumping water from the bottom of the plastic container to the top, so that it constantly flows down through the material. As the smoke passes through the flue gas scrubber, the polluting particles found in the smoke settle on the moist material before the smoke finally leaves the house. The water from the bottom of the plastic container gradually becomes filled with nutrients and can be used as liquid fertilizer.[12]

Another more standardized solution is to install an electrostatic stove filter.[13] The stove filter is installed at the top of the chimney. Testing of such a filter has shown that it reduces the number of fine and ultrafine particles by 95% and reduces the total particle mass by 70–75%. The filter also seems to have a good effect on the emission of black carbon.[14] The stove filter does not have as many useful functions as a flue gas scrubber, but is easy to install if you already have a stove.

Local cogeneration

With cogeneration, electricity and heat are produced simultaneously and can use local biomass. Cogeneration is an energy-efficient solution because when we burn wood to produce electricity, there will inevitably be a loss of energy in the form of heat.

This heat is used for home heating, hot water or similar. Unlike electricity production from the sun and wind, the cogeneration plant supplies electricity when we need it, regardless of wind and weather. The gasifier and the stirling engine are two technologies that can be used for local cogeneration, where wood is burned, for example.

Gasifiers

A gasifier is a type of furnace that produces flue gas by burning wood by pyrolysis. The gas is cleaned through various filters and can then be used as fuel in a slightly converted gasoline engine that drives an electric generator.

Both the gasifier and the engine become hot during operation, and ideally this heat is used. The ratio is two parts heat to one part electricity,[15] and degassing 1 kg of wood can produce 1 kWh of electricity, and on top of this comes the production of heat. In some gasifiers, it is also possible to retain a smaller portion of the carbon in the biomass as biochar, so that the overall energy production becomes carbon negative.[16]

The challenge with gasifiers is that there can easily be tar residues in the flue gas, which can damage the engine. One way to get around this problem is to produce charcoal in a biochar stove and use the coal as fuel in the gasifier.[17] Charcoal is completely pure carbon without tar. If you burn your charcoal, you obviously do not achieve carbon storage, but on the other hand, you do produce your own fuel.

Stirling engines

Stirling engines and organic rankine cycle engines can also be used for cogeneration plants. These are heat engines, i.e. it is a temperature difference that makes them run.

They can be powered by any heat source, both solar heat and the burning of biomass. It is therefore possible to connect a stove that produces biochar, e.g. a TLUD stove. Here, about five to ten parts heat is produced to one part electricity.[18]

Solar cells and wind turbines

Solar cells and wind turbines provide us with electricity with a much lower climate impact than coal power, but they are still made from non-biological materials, and their production involves the consumption of fossil fuels and other raw materials that burden the environment and climate.

Solar cells have a very varying climate impact during production. Solar panels manufactured in China have a much greater climate impact than solar cells from Europe. This is because production requires large amounts of electricity, and the Chinese electricity grid is very CO_2-intensive.[19, 20] Today, more than 80% of the world's solar cell production takes place in China.

Another factor that has an impact on the climate impact is the type of solar cells. Monocrystalline solar cells, the most common type on buildings, usually have a greater climate impact during production than polycrystalline solar cells and thin-film solar cells.[21]

On south-facing roof surfaces, we can use solar panels that also function as roof tiles. This saves resources that would otherwise be spent on another roofing material. This is a good example of the principle that one element has multiple functions. By placing solar cells on roofs instead of on racks on the ground, we also avoid using land for this purpose.

Wind turbines generally work best on a large scale, because wind conditions are best at high altitudes. A tower 25 meters or more high is usually necessary to reach where there is good wind without turbulence.[22] Harvesting wind energy also often conflicts with our desire to improve the microclimate by creating shelter. However, if we are off-grid, a small production of electricity from a wind turbine can be important, especially in winter.

Airborne wind energy systems are currently being developed. These consist of various types of kites that produce energy, for example by the kite pulling on a tether causing a drum to rotate, which in turn drives a generator. An advantage of these systems is that you avoid having to build a mast to hold the turbine, and the investment of embodied energy is much less per kilowatt hour than with ordinary wind turbines.[23]

Solar cells and wind turbines harvest energy depending on the weather. During the winter months we may experience periods where there is neither sun nor wind. Therefore, it is a good idea to use cogeneration as a backup. If we focus on solar and wind, we can avoid overconsumption of biomass.

Hydropower

Like cogeneration plants, hydropower can provide us with a stable energy supply. Large-scale hydropower often has major environmental consequences, and the possibilities for establishing such plants are fully exploited in most places. If local conditions allow it, small-scale hydropower is a good solution that can secure the electricity supply in a permaculture project.

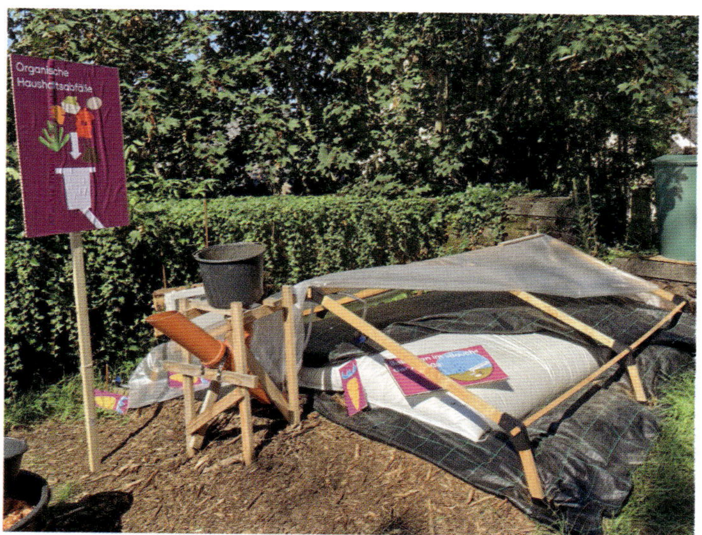

The biogas plant at Katrin Pütz's in Germany produces gas for cooking. The digester is covered by a greenhouse, that keeps it warm three quarters of the year. In the front of the picture you can see the inlet, which is used for adding food waste.

BOTH: KATRIN PÜTZ

Biogas can be stored and transported in a plastic bag. Katrin has started the company (B)energy, which produces biogas plants that are sold in both Africa and Europe.

Household biogas plants

Biogas plants at the household level are a widespread technology in India, China and parts of Africa. A biogas plant is built around a container with water, which is filled with organic material such as manure, food waste, vegetable scraps, leaves and grass. Microorganisms that are adapted to oxygen-free conditions convert the material and release methane gas (CH_4) that bubbles up from the tank. This gas is collected and stored in another container, which in small plants is usually a large plastic bag.

Every time organic material is put into the container, fertilizer water flows out of a pipe at the other end, and this can be used for cultivation. The fertilizer value of the materials that we put into the digester is maintained, but the carbon content (C) is reduced, as biogas (CH_4) contains carbon. The biogas from small plants is usually used for gas stoves, but if the production is large enough, the gas can also be used to drive a generator and produce electricity, but this requires the engine to be adapted to biogas. Under optimal conditions, a $1m^3$ biogas digester can be fed with 10 liters of food waste per day, thereby producing $1m^3$ of uncompressed biogas. The energy content of $1m^3$ biogas corresponds to 0.5 liters of oil and can be converted into 1.7 kWh of electricity if it is burned in a generator plant.[24]

By building small biogas plants instead of large ones, the transport distances for the material to be degassed are reduced, and in practice some material is utilized that would not have been used for anything if there were no plant on site.

In warmer climates, the soil temperature is so high that, simply by burying the biogas plant, an optimal climate can be achieved for the microorganisms that ensure the development of biogas. This poses a challenge in relation to small biogas plants in temperate climates.

5

Construction

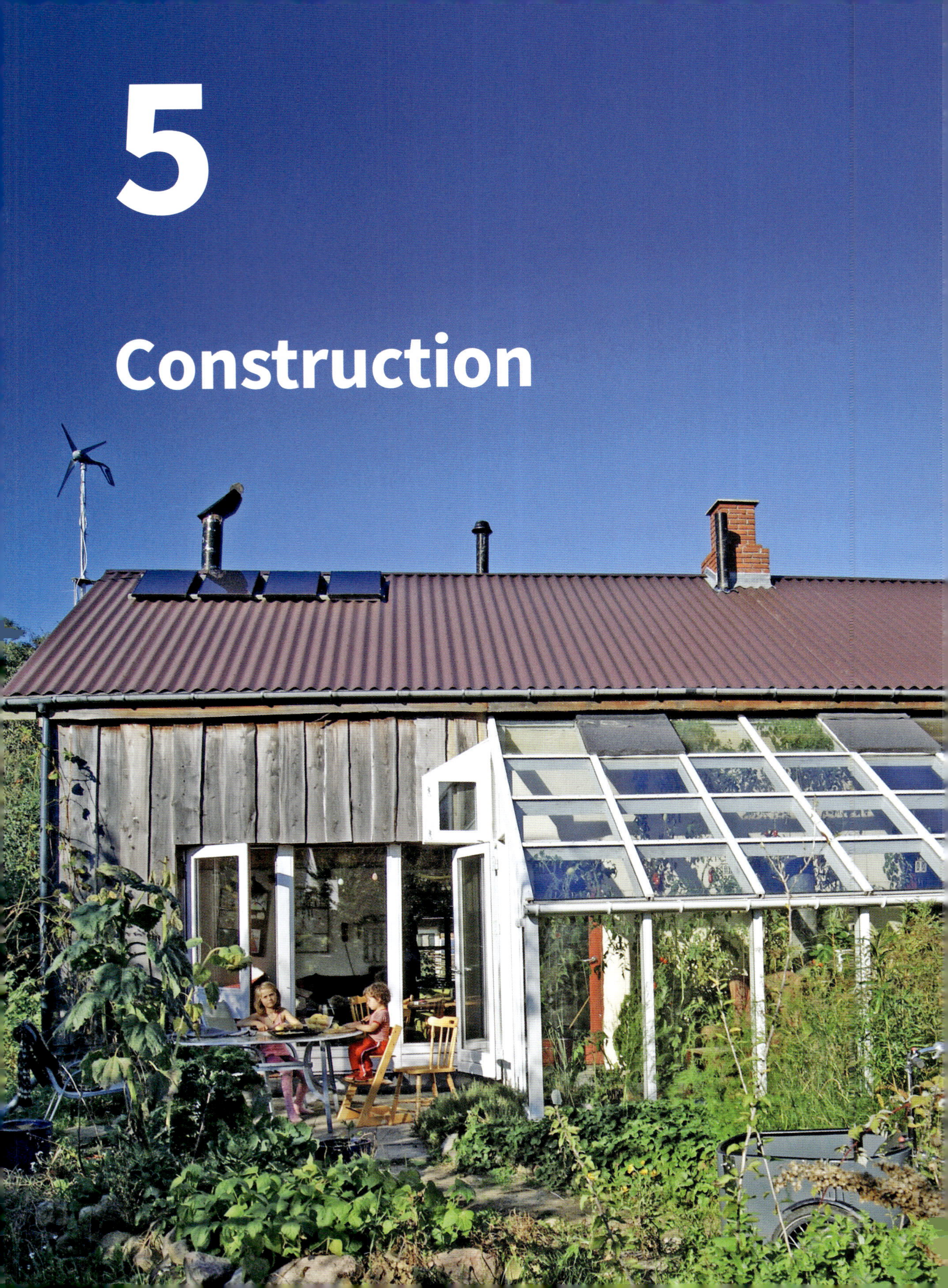

The vision of low-energy houses of local natural materials

The vision within permaculture construction is to build houses that help regenerate natural resources and mitigate climate change: well-insulated houses that do not need fossil fuels for heating and electricity, but instead use local resources such as sun, wind, wood and other biomass. These houses are built from local natural materials, many of which will store carbon throughout the life of the house.

In temperate climates, the house is a large and important part of a permaculture design. We spend many hours at home, with different rooms serving numerous important functions in daily life, and this is where we use a lot of our natural resources. In hot climates, the house is where we can seek shade and coolness.

Principles of construction

When it comes to construction, the use of the primeval forest as a model is not as direct as in cultivation systems. But many of the permaculture principles that can be related to the primeval forest are still in play.

For example, we aim to use *biological resources* made by photosynthesis, such as wood and straw, rather than concrete, steel and rock wool. This will result in low embodied energy consumption; at the same time the house must be well insulated and well designed so that it *harvests, stores and utilizes energy flows.*

We aim for the house to have *multiple functions* by being careful with the *relative location* of the elements in the house. Besides the main function of being our residence, the house can also be a carbon store and produce electricity, heat, compost, food etc.

The *edges* between the house and its surroundings need to be productive with cultivation and rainwater collection, and instead of standard homes, there must be a sense of place, so that the house is adapted to local materials, climate conditions and the people who live there.

The house is designed to be part of a *whole,* in a cycle with its surroundings, so that all outputs become inputs somewhere else.

The straw-bale house in The Self-Sustaining Village has many functions. In addition to being a home and workplace, the house produces food in the greenhouse and compost from the kitchen and toilet, electricity from solar cells and a wind turbine, heat from passive solar heating and a rocket stove mass heater and hot water from the solar collector on the roof of the greenhouse. Finally, the house stores 22 tonnes of CO_2 during its lifetime. Built by Esben Schultz and Mira Illeris, Denmark.[1]

The flow-through house

Most houses today receive a lot of natural resources in the form of clean drinking water, nutrients contained in food that is purchased, plastic products, energy in the form of gas, oil and electricity, etc. These natural resources come out of the house again in the form of various types of waste such as wastewater, CO_2 and mixed waste, which, depending on the waste disposal system you have, is sorted into several categories, and a small part of it is possibly recycled.

Waste is transported with an associated energy consumption away from the vicinity of the house. The natural resources are broken down.

The image of the flow-through house can also be used for a local community or a city, and the vision of permaculture is to create closed cycles where all outputs

Old half-timbered house renovated with additional insulation, greenhouse and solar collector on the south side, compost toilet and masonry stove. Bjarne Wickstrøm, Denmark.

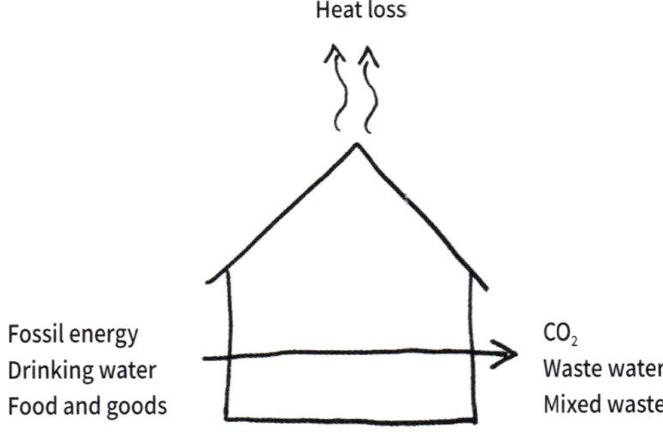

Nutrients, minerals and water circulate in closed loops between the house and its surroundings in the 'circular house', while in the 'flow-through' house it comes out as waste. There will inevitably be heat loss from the house, but we must strive to ensure that the house has the lowest possible need for energy for heating and cooking, etc. and ensure that the energy comes from renewable energy sources.

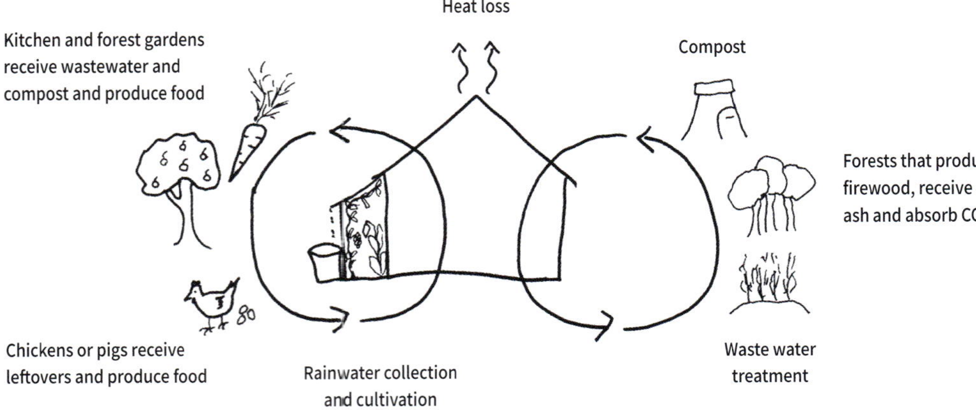

The green home in Vrads was a completely ordinary brick flow-through house. Today it is a circular house. It has been insulated, solar collectors have been installed, a compost toilet and a masonry stove have been built, and there is rainwater in the taps. The edge has been made lush with greenhouses and climbing plants, and it all forms part of a cycle with the kitchen garden and the forest behind the house. Denmark.

become inputs somewhere else in the system and there is no waste.

The circular house

The circular house is part of a coherent whole with its surroundings and people; the house and its immediate surroundings together build up the natural resources. The house receives water, food and firewood from the land around it, including local agriculture, and the house delivers nutrients and carbon from kitchen compost, the compost toilet, ash and wastewater back to the land.

The house captures the sun's energy and produces heat and electricity. It collects rainwater for irrigation and washing and has a lush edge, e.g. grape plants or climbing beans, which grow up the south facade and produce food, but also more abstractly with glass and solar cells that harvest the sun's energy.

Resilience

The house, together with its immediate surroundings and its local area, is resilient to any challenges that climate change may bring and to failures in external supplies of food, electricity or heat. The house can be repaired from local materials and with local knowledge. The house is a central piece in a permaculture lifestyle.

KAROLINE NOLSØ AAEN

Lushness around and up the strawbale house at Myrrhis Agroforestry in Denmark. Caucasian spinach grows up the extension on the north side of the house, and these climbing plants make use of an otherwise unused cultivation area. In the roof garden on the roof extension, potatoes and pumpkins are grown, among other things.

House design

The following section will provide solutions on how we can optimize our house design without extra work and expense, solutions that can be combined into many different versions of permaculture construction. Few of us choose to build a new house from scratch, but the design ingredients can also be used in connection with remodeling, renovation and house purchases.

The house is our zone 0, the most intensive zone, where we spend the most time and where we therefore make the most of the microclimate. If we can choose where to build the house, it should therefore be located in the warmest place in the landscape, and we will, as far as possible, improve the microclimate around the house by creating shelter and warmth. The house must be kept warm, and much of the task of designing a house in a temperate climate is about how we harvest energy for heating and keep the heat in the house.

The small house

Limiting our consumption and living in a small house with fewer square meters is the cheapest and simplest step towards a design with low embodied energy and low ongoing energy consumption. It is an important part of living the permaculture ethic of caring for the earth and fair share.

MARIE COX

'The small potential' is a 20m² tiny house for two people. It was built in six months from cheap recycled and natural materials such as wood, wood-fiber insulation and clay plaster. It is designed with a focus on space utilization, functionality, water collection and the possibility of being both on- and off-grid. Solar energy can be harvested and stored. Denmark.

As a general trend, the living space per person has increased over the past decades in northern Europe and the USA,[2,3,4] and newly built homes have a larger area than before. In some of the world's rich countries, the average living space is now between 40m² and 90m² per person.[5,6] The increase is due to several factors, such as more people living as singles and the fact that with increasing incomes we want to live in larger homes. In comparison, the research report 'A just world on a safe planet' defined the minimum number of square meters for a life without poverty as 15m² per person.[7] The living space per person is very unevenly distributed, so despite the general increase in living space, there are also people, and especially families with children, who live in overcrowded homes.

A Danish study shows that the increase in living space means that, although energy consumption in new buildings has fallen significantly per square metre since the 1970s due to increased insulation thicknesses and other efficiency

The straw-bale house at Reforest Farm with kitchen–family room and three smaller rooms is 68m² and is a home and workplace for two adults and two children. The house is recessed into the south-facing slope so that it takes advantage of the root cellar effect, and has a planted roof so that it blends into the landscape. It is off-grid with passive solar heating, wastewater greenhouse and compost toilet. Denmark.

improvements, total energy consumption for heating has remained at a constant level, simply because buildings are being built larger and people live in more and more space. Calculations for CO_2 emissions from new buildings are made in CO_2e per square meter per year and do not look at the total living space. There is therefore a lack of focus on the problem of overconsumption of living space.[8]

Perhaps we should start asking ourselves whether a large home is the way to a good life? Or whether it is perhaps better to live in a more modest living space, if it allows us to have less debt and more time in everyday life?

Compact house with minimal building envelope

It is a good idea to design the house so that it has the smallest possible surface, the smallest possible building envelope. This will result in less heat loss. Round houses and domes are examples of how the building envelope can be minimized on smaller buildings. However, it should be noted that with straight or square materials, e.g. boards, beams and straw bales, it is much simpler and thus more labor-saving to build a simple square house with a two-sided sloping roof than to build houses with different forms of organic design.

Another way to minimize the building envelope of our homes is by having multiple housing units in the same building. The energy consumption related to heating is around 30% higher for a detached house than for a similarly insulated terraced house of the same size, based on houses with a moderate level of insulation.[9] If instead we are talking about low-energy houses with heat pumps or district heating, which get their energy from an energy mix with a high proportion of renewable energy, the difference in climate impact between the detached house and the terraced house is not very large.[10]

MICK HART

Compared to detached houses, terraced houses have a compact climate shield, which saves materials and reduces heat loss. Munksøgård Community, Denmark.

In Bjarne Wickstrøm's cob house, passive solar heat is stored by the sun's rays hitting heavy cob walls and a clay floor. Denmark.

Passive solar heating

Passive solar heating is the heat we get when the sun shines through the window. This form of solar heating is called passive because it does not use mechanical or electrical parts, unlike, for example, a photovoltaic solar collector. It means we can capture the sun's energy with very little financial cost and a small consumption of embodied energy, whether it is designed into a building from the beginning or when renovating an existing one. Skylights for passive solar heating installed in a renovation project can deliver energy at a cost per kilowatt-hour that is comparable to the cost of electricity generated by solar cells,[11] and the energy from passive solar heating and that from solar cells can actually complement each other because the usable heat energy from passive solar heating is delivered in the winter months, while energy production from the solar cells is greatest in the summer months. In a house where solar cells and a heat pump are installed, a greater degree of self-sufficiency in energy can be achieved by also equipping the home with passive solar heating.

Passive solar heating can make a significant contribution to a home's heating needs in most locations, but it will usually still be necessary to have another source of heat to fall back on during dark and cold periods.

Window orientation

Research in sunny, dry climates in the 1970s and 1980s suggested that passive solar design should focus on vertically oriented south-facing windows. New research is now being conducted by Alexandra Rempel and her collaborators at the University of Oregon. This research shows that better performance can be achieved in almost all climates, especially those with significant winter cloud cover, by using tilted windows that are oriented to the south, southeast or southwest (in the northern hemisphere).[12] Skylights are particularly good because building roofs are usually not shaded by trees and other nearby buildings. Vertically oriented windows are, according to the new research, only ideal for harvesting solar energy in very hot and sunny climates.

In the temperate climate zone, cloudy weather during the part of the year when we heat our homes is the norm rather than the exception, especially near the coast, and a large part of the solar energy available is therefore diffuse sunlight.

When it is completely or partially cloudy, the clouds scatter both visible sunlight and infrared light energy towards the top of the skydome, and the solar energy will then come from a higher area in the sky than the position of the sun.

In almost all climates, the best performance of passive solar heating can be achieved through a design with tilted, south-facing windows. Such a design is optimal for capturing the diffuse sunlight that comes from a higher area in the sky than the sun's position. At the same time as maximizing the capture of solar energy, it is also a good idea to reduce heat loss through the windows at night by using movable insulation.

Therefore, tilted south-facing windows are ideal for capturing the diffuse light.

At the beginning and end of the heating season, i.e. October, March and April, there is a lot of direct solar radiation in most places, and the sun is relatively high in the sky. This sunlight is also captured more effectively by tilted south-facing windows than by vertical windows, and during these months it is a particular advantage to have tilted windows.

The optimal angle of the windows is 45–60 degrees above horizontal in most of the United States, but in some cases, the optimal angle is as low as 30 degrees above horizontal. The precise angle is calculated based on the geographical location and depends, among other things, on the degree of cloud cover. The cloudier the climate, the further away from vertical the windows must be to capture the diffuse solar radiation.[13]

Windows facing east and west also receive sunlight but mainly in the summer months at the beginning and end of the day, when we do not need the heat and want to avoid the house overheating. Other needs can impact our choice to have windows that are not south facing: the desire to get daylight into the house or that there should be a view.

The guidelines described here for the orientation of windows in relation to harvesting solar energy apply not only to passive solar heating design in the home but also to attached greenhouses and conservatories, where tilted south-facing windows can also be advantageously incorporated.

Movable insulation

Even the best low-energy windows are less well insulated than an insulated external wall, and windows are therefore one of the biggest causes of heat loss from buildings.

In order to achieve a large overall heat gain from passive solar heating, it is therefore important to have movable insulation that can reduce heat loss through the windows in the morning, evening and night. This applies to both the tilted windows that are designed to receive passive solar heat and the other windows. The movable insulation is ideally removed a few hours after sunrise and pushed back again one to two hours before sunset, because the heat loss from the window in the morning and evening is greater than the heat gain. However, movable insulation is also beneficial even if you uncover the windows in the morning to get a view. An advantage of skylights is that movable insulation can be used without disturbing the view from the house.[14,15]

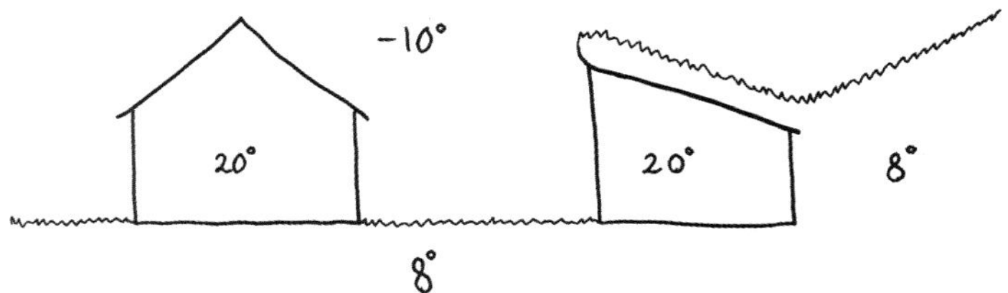

The difference between the room temperature and the outside air is 30°C, while the difference to the soil temperature is only 12°C.

Shade curtains

Shade curtains are used to provide shade and thus avoid overheating in the summer.

It has previously been considered good design to prevent overheating of buildings by creating a wide overhang over the windows that are to receive passive solar heat so that the solar radiation is shielded in the summer, when the sun is high in the sky. However, such a wide overhang blocks out diffuse sunlight during the heating season. A better design is to install shade curtains, which at all times provide the possibility of stopping the solar radiation. The shade curtain is drawn when unwanted solar radiation comes in through the window. During hot periods however, it is an advantage to have it open at night, to maximize the radiation of heat out of the window.[16]

Storage of passive solar heat

Passive solar heat can be advantageously designed so the sunlight hits a heat-storing material that absorbs the solar energy and then slowly releases radiant heat during the night. This could be, for example, a dark tiled floor or an adobe wall.

A large untapped resource

Calculations have shown that if 10m^2 of south-facing skylights were installed in every home in the United States, it could reduce the nation's energy needs for home heating by as much as 30%, so there is a lot of energy to be gained from passive solar heating. There is also a lot of energy to be gained from passive solar heating in cold and cloudy climates.[17]

The root cellar effect

Due to the enormous mass of the earth, which slows down temperature changes, the temperature two to three meters below the ground is relatively stable all year round and corresponds to the average temperature for the whole year. This has been used for centuries when building root cellars, but this fact is also important when we design homes. The part of the house's building envelope that faces the ground rather than the outdoor climate will have less heat loss in winter, as the temperature difference will be smaller. The root cellar effect is exploited when houses are placed with ground contact rather than on raised point foundations. We can also choose to build the house partially into an earth embankment or a south-facing slope and reduce heat loss even more.

Add a conservatory or shed

By placing a conservatory or a greenhouse (if the main purpose is growing plants) as a south-facing extension to the house, many useful connections are achieved. Both the house and the conservatory will have less heat loss than if they were located separately. The conservatory will reduce wind cooling of the house and harvest passive solar heat, which can be let into the house during the day. In return, the conservatory will receive the house's heat loss at night for the benefit of the plants.

To get the warm air from the conservatory into the house, the conservatory is designed so that in winter cold air can be taken in at the bottom, which when heated will rise upwards and from here can be taken into the house via a ventilation hole or a window. Alternatively, the air can be taken into the conservatory from a buried pipe so that it is already preheated by the ground before it enters the conservatory.

If it is not possible to place the conservatory facing south, it is best to place it facing southeast, where the morning sun will warm the house after the night's cooling. Facing southwest, it will add heat to the house at the end of the day, when the house is already warm, with the risk of overheating. If the conservatory is placed facing southwest, it is therefore extra important to use heavy materials where the solar heat can be stored.

The placement of the conservatory as an extension to the house gives the house an extra and very special room – in line with the principle of *edge effect*, you could say that the conservatory is a lush edge to the house. When the conservatory is located on the edge between the house and the garden, it can be called zone 0.5. It is practical to be able to go directly from the house into the conservatory, and the plants will be looked after more often.

CATHRINE DOLLERIS

This straw-bale house in Birkegården gardens has a greenhouse to the south and an outhouse to the north. Denmark.

By also placing a shed or outbuilding as an extension on the north side of the house, we reduce wind chill and heat loss and achieve a compact shape. Storage of food, materials and tools used in zones 0 and 1 is well placed here.

Zoning the house

One way to limit the need for heating is to place the most used rooms, such as the living room and kitchen, to the south, where they can be kept warm by passive solar heat for a large part of the year, while bedroom, toilet, hallway, entrance hall and utility room are placed to the north. Such a location will also save on electricity consumption, as the most used rooms during the day will then receive the most natural lighting from outside.

Optimize the existing home

Sheds and greenhouses can be added to existing homes. If the building is L or U shaped it might be possible to place the building extension in such a way that it reduces the cooling of not just one but two or three external walls.

In an apartment, you can make a small garden cloche outside a window. By installing a balcony box with plants in it and putting glass around it, we will be able to open the window on to a small intensively cultivated bed, get preheated air into the apartment and harvest fresh greens.

We can also create sensible zoning when we choose what different rooms will be used for, and whether any of them may not need heating to ideal room temperature.

However, partition walls and interior doors are usually poorly insulated, and if there is a large temperature difference between the rooms, the partition wall and door can become a relatively cold surface in the warm room, where condensation can occur. To avoid mold, the temperature difference between rooms should therefore not be more than 5°C and no room should be below 18°C.

The root cellar

The root cellar is an important design element in temperate climates. The root cellar is a cool storage room that is built into the ground on five sides and has a door that we can enter through. If there is already a basement in the home, it can also be a good solution to arrange a corner of it so that the same conditions are achieved as in a root cellar. In the root cellar, we can use the stable temperature underground to achieve cool but frost-free conditions without the use of

RALF PALMPERS

The root cellar is located right next to the kitchen garden, so it is easy to move the vegetables for storage. Växhuset, Sweden.

JAN GUSTAFSON-BERGE

The root cellar keeps the cold out in winter, Lövudden, Sweden.

electricity, thereby enabling the storage of large quantities of fruit and vegetables during the cold season, when there is not much to harvest outside.

A root cellar can also be used to store juice, wine and jam, and shrubs and perennials that are not winter hardy can be dug up and overwintered here. There must be suitably high humidity in the cellar so that the vegetables do not dry out, but not so high that condensation occurs. An earthen floor can help to release moisture into the room.

It is a good idea to make ventilation with air intake at the bottom at one end and an air outlet at the top at the other end. If it is difficult to keep the temperature down in the fall and spring, it may be necessary to only have the ventilation holes open at night, when the air is coldest. Another option is to let the air in via a buried pipe, which will moderate the temperature of the ventilation air before it reaches the cellar. Shelves are made with a distance of 5cm from the wall, which allows ventilation air to pass behind them.

It may also be a good idea to make the cellar with compartments so that fruit and vegetables can be stored separately, as fruit releases ethylene, which accelerates the aging process of vegetables.

The cellar is important in a household where one wants to work seriously towards self-sufficiency and in a permaculture farm. It should be placed in zone 1, so that it is easily accessible for daily cooking.

Composting toilet

The flush toilet has several negative aspects. Clean drinking water is destroyed and turns into a difficult-to-handle sludge of nutrients, pathogens and heavy metals. Large amounts of energy are used to first provide the clean water and later to clean the sludge at municipal sewage treatment plants so that, for example, the nitrogen is converted to a gaseous form. At the same time, energy is used to extract nitrogen from the air for fertilizer in factories.[18] Establishing a composting toilet helps to bring the home into circulation with the local surroundings, and it turns waste into a valuable resource.

Compost toilets can be constructed in different ways, and the following is based on our own experiences of what works best. A modern compost toilet has urine separation, where the urine is collected in a funnel and directed via a pipe to a tank, and the feces are kept dry and sprinkled with wood shavings or similar after using the toilet. The chamber in which the feces container is located is ventilated out of the room, so that the toilet is completely free of odors. The chamber is also made so that the feces container can be taken out via a door in the outer wall of the house. The feces container should not be larger than we can carry when full, for example a 30-liter bucket. The actual composting takes place in two outdoor compost bins. There should be two, so that one can be filled while the other is composting without the addition of fresh material. It is important that the compost bins are ratproof.

Fresh urine does not contain dangerous bacteria in principle, but bacteria will form when it is stored. One option is to use the urine under fruit or nut trees that are not underplanted with an edible crop so that the urine does not come into contact with the parts of a plant that will soon be eaten. If we use the urine in cloudy weather and immediately before rain, minimal nitrogen will evaporate. The rain will water down the urine and this also reduces odors. A permanent root system from trees and other perennial plants also ensures that leaching does not occur.[19]

A single person produces about 500 liters of urine per year. One liter of urine contains about 8g of nitrogen.[20]

AGNER THULESEN

The compost toilet in det grønne hjem (the green home) in Vrads, Denmark.

The amount of urine from one person is enough to fertilize about 300m^2 of apple or pear trees, or to fertilize about 500m^2 of hazel.[21]

Composted feces are much safer to use than sewage sludge from a sewage system where many wastewater sources are mixed together, but we should still adhere to the precautionary principle and only use the fecal compost where food is not produced. One should also be aware of the legislation on the subject in the country in which one lives.

Composted feces are rich in potassium and phosphorus but do not contain much nitrogen. It may therefore be a good idea to use them to fertilize nitrogen-fixing plants, or possibly a polyculture that includes nitrogen-fixing plants. A coppice with common alder or false acacia could be an option. The phosphorus content in feces from one person is suitable for fertilizing approximately 500m^2 of common alder.[22,23]

Building materials

The climate impact of building new

Throughout the last 30–40 years, there has been a focus on reducing the day-to-day energy consumption in our homes. However, this development has a downside because it often increases the consumption of materials and technical installations that emit large amounts of CO_2 during production. When calculated according to a building life cycle of 50 years, the materials required to build a new low-energy house today account for approximately 75% of its total greenhouse gas emissions, while the operational energy consumption accounts for 25%.[24]

Modern building materials such as cement, steel, glass, aluminum and mineral wool each have a high embodied energy. Cement alone accounts for 8% of humanity's CO_2 emissions.[25] This is due not only to the energy consumed in production, but also the chemical processes that occur when the cement hardens.

Building a new low-energy house out of industrial materials results in a climate impact of 400–600kg of CO_2 per square meter.[26,27] That is around 50 tonnes of CO_2 for a home of 100m^2.

Emissions from the production and transport of materials occur long before move-in day. The fact that a new building saves energy over time is not a sufficient justification for high energy consumption in the here and now, as climate change is an urgent problem which requires society to reduce greenhouse gas emissions not only in the long term, but also in the short term. This is a crucial factor in circumventing climate tipping points.

Recent climate research has introduced the concept of the 'carbon budget', which refers to the remaining amount of greenhouse gases that the atmosphere is expected to be able to absorb while maintaining a global temperature increase of less than 1.5°C. If we compare the yearly carbon budget allocated to an individual for housing to the carbon footprint of a new, conventional, four-person single-family home over its lifetime, it emerges that the CO_2 emissions of the build are approximately 10 times the budget of each person in the household. This is calculated over the expected 120-year lifecycle of a building, so in fact, the occupants of a newly built house will overshoot their annual CO_2 budget for the home by 10 times for every year they live there.[28,29]

Instead of erecting new low-energy houses out of industrial materials, it is often better to mitigate climate change by renovating existing buildings. This utilizes the embodied energy that has already been invested.

Local natural materials

We must prioritize the use of locally sourced natural building materials when renovating and when building from scratch.

Energy consumed in transportation can be minimized if materials that are large and heavy, and those that are required in significant quantities in construction, are sourced locally. It is possible to build low-energy houses that are composed chiefly of more or less unprocessed natural materials such as clay, sand, stone, seashells, wood, straw, eelgrass and reeds.

Example of CO_2 emissions over the 50-year life cycle of a newly constructed home. Note that emissions from materials are relatively high during both construction and disposal. Overall, materials account for 75% of CO_2 emissions while operations (heating, hot water and ventilation) make up the remaining 25%. As such, the use of natural and recycled materials should be prioritized.

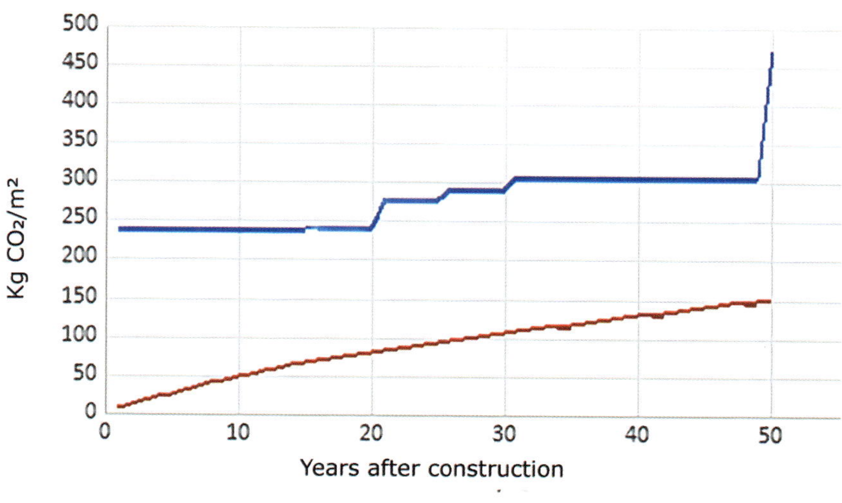

FREJA RASMUSSEN

Not only does the use of these materials save energy and mitigate climate pollution, it can even counteract climate change, as building materials such as straw and wood store carbon throughout the lifetime of the house.

An analysis of the single-family straw-bale house depicted on page 162 revealed that the straw stores carbon equivalent to 22 tonnes of CO_2, while other materials in the same building and their transport etc. resulted in emissions of 10 tonnes of CO_2, leading to a net storage of 12 tonnes of CO_2.[30]

By using local materials, we are also strengthening the local economy where we would otherwise transfer wealth to multinational corporations.

Hazardous substances

By using natural materials we avoid the hazardous materials that are present in large quantities in industrial construction, especially filler materials, paint and glue. These are substances that can cause eczema, allergies, lung disease, poisoning and cancer. Houses built from natural materials on the other hand, can easily be returned to nature when they are demolished.

Can sustainable construction be upscaled?

Materials such as timber and recycled building material emit a low amount of CO_2 for a single building, but it is difficult to scale this up as a viable alternative to industrial construction. Recycled materials would simply be impossible to source in sufficient quantities, while upscaling the use of timber could easily lead to unsustainable forestry and deforestation. Thus, we conclude once again, we must reduce the size of our living spaces.

There are some potential exceptions, such as straw production in grain-producing areas. Upscaling would be possible as long as we grow as much grain as we do today. Similarly, the use of eelgrass could also be scaled up and included in a regenerative context while there is an excess of nutrients in the oceans.

The timber used for construction at Skovvirke comes from the local forest. It is sawn at the community's own sawmill and dried in the open air. Industrial timber found in hardware stores is typically kiln-dried. This process accounts for 70% of its embodied energy.

Clay soil is a material that unlike wood, is abundant globally. Unburnt clay can even be recycled.

Today, in many countries there are not enough locally produced reeds for thatching, and reeds end up being imported from distant places; Europe sources some of its reed from China. However, climate change makes it necessary for wetlands to be recreated in places where there are currently carbon-rich drained agricultural soils. This will increase the production of reeds.

Natural industrial materials

Lime mortar and insulation materials made from fibers of wood, hemp and flax are among the natural materials that are often utilized in ecological building. However these materials are produced industrially and are rarely obtained locally.

Wood fiber insulation batts are one of several examples of manufacturers embracing natural materials, thus making natural building materials more accessible to the public. Upscaling production is not necessarily feasible, however, as wood is a finite resource.

JETTE KRUSE LARSEN

Jette has planted sedums on the green roof of her straw-bale house in Denmark. These plants came from the local environment and therefore support the insect life and biodiversity of the area.

Recycled materials like bricks, roof tiles, windows and doors are also used in ecological building and contributes to the reduction of embodied energy.

Roofing materials

Thatched roofs and shingle roofs are the most sustainable that can be produced locally in most places. Thatched roofs have an insulating effect, and this can be augmented by adding flax fiber insulation batts on the inside.

Steel sheets and roof tiles both consist of materials that consume a lot of energy in the manufacturing process. However, since the roof is only a thin layer, the embodied energy of these materials is not unreasonable when viewed in relation to the total energy consumed in the house's construction. These roof types are among the best for rainwater harvesting.

Green roofs, or living roofs, are heavy because the soil is heavy, especially when wet. This places greater demands on the strength of the structure and increases material requirements. The weight could be reduced with the use of sedum plants, which can grow in just 5cm of soil. Sedum roofs are also more fire-resistant than grass roofs, which can dry out and become flammable in the summer. The advantage of green roofs is that they are long-lasting because the roofing material is protected by soil, and therefore not exposed to the sun's UV radiation. Green roofing allows buildings to blend in with nature. In cities, living roofs can provide cooling during heat waves and support biodiversity in places where there are few habitats for insects and other wildlife.

KARIN MALMGREN

Wood roofing shingles can be made locally. Here the material is being installed on a building at Ecotopia in Sweden.

Materials for walls and insulation

Buildings made from natural materials are usually constructed as diffusion-open structures without a vapor barrier. This means that moisture can move through the materials and stabilize the humidity level in the buildings, which helps to create a healthy indoor environment. The fact that a building is diffusion-open does not mean that it should be left susceptible to air leakage as a result of wind, as this would increase heat loss enormously and increase the risk of condensation inside the insulation materials. Buildings are made airtight so we can control when and how much ventilation is needed.

Straw bales, eelgrass and reeds are insulation materials we can most often source locally. Walls and ceilings insulated with straw are plastered with a clay plaster to make them windproof and fireproof.

Clay plaster can generally be extracted locally, and potentially even dug up where the house is to be built. Sand or clay may need to be added to achieve the right composition. If a wall or a ceiling made of planks is to be plastered, a good plastering surface can be attained by installing reed mats, i.e. interwoven mats of reeds. Clay plaster and lime mortar have the advantage over cement in that they are flexible materials that can move without cracking when there are small subterranean movements. Therefore, depending on the construction of the building, it is possible to make very durable houses from these materials. In situations where clay rendered exterior walls are expected to be exposed to driving rain, it may be necessary to use cladding, for example untreated larch boards, to protect the render from the elements.

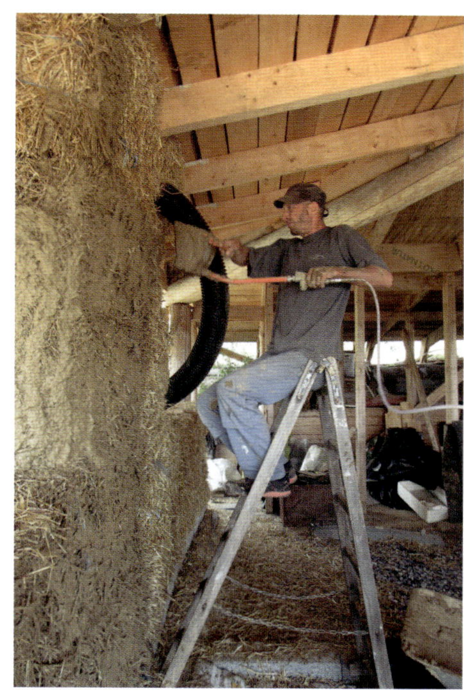

Straw-bale walls are plastered with clay to make them airtight and fireproof. Skovvirke, Denmark.

Clay walls can be made either with adobe, i.e. dried and possibly pressed bricks made of sand, clay and in some cases straw, or by using cob, where clay, sand, and straw are kneaded together to form large loaves that are then used to build the walls. Building with cob is relatively time-consuming, but the materials can be fully or partially sourced on site. Cob is a material that offers great freedom of shape, and you can build almost without the use of machinery. Heavy adobe and cob walls should be placed on the inside of the insulation where they can act as a thermal store, helping to even out the temperature differences between day and night. Thin inner walls can be built from cob on a frame of braided branches.

BJARNE WICKSTRØM

With cob, organic shapes and forms emerge naturally during construction.

Straw-clay is an old building technique that can be used for exterior walls, but is perhaps especially advantageous for interior walls as they can be made thin. Loose straw is mixed with slurry clay, which has the consistency of yogurt, and rammed into a movable formwork between vertical battens. Cross braces are

KARIN MALMGREN

Sun-dried unburned adobe bricks made from local materials can be produced entirely without the use of man-made energy. Ecotopia, Sweden.

The cob loaves are moulded and ready for use.

BJARNE WICKSTRØM

placed every half meter to stabilize. A wall can be erected in a day.[31]

Charcoal as an insulation material has been trialled by the Ithaka Institute in Switzerland. Charcoal was mixed into clay plasters in a ratio of 50% charcoal, 30% gravel and 20% clay.[32] According to the institute, a plaster such as this has several advantages. It is thermally insulating because the charcoal has many small cavities, and it is even better than regular clay plaster at stabilizing the humidity in a building and purifying the air. Additionally, the building will be a long-term carbon store even after the building is demolished and the clay plaster is returned to the ground.

Materials for foundations

When constructing the foundation of a building, we must seek to minimize the use of concrete.

A foundation must be made, of course, to last: once the house is built, it can't be replaced. Professional straw-bale house builder, Barbara Jones, describes in her book *Building with Straw Bales*[33] various alternative methods for the construction of foundations, citing how we often oversize them. In ecological building, for example, pier foundations are often used instead of perimeter foundations, where the foundation is poured only under the load-bearing posts. This saves large amounts of concrete.

Alternatives to cement

In the past, boulders were used to distribute the weight of posts and walls. Sometimes rubble trench foundations were used, where smaller stones were rammed into a hole or trench. We can still do this today.

It may be necessary, however, to use more advanced materials to construct a foundation that is free of cold bridges.

The outer straw-bale walls and straw-clay partitions are being erected. The reed mats in the ceiling are still visible, yet to be plastered at this stage. Skovvirke, Denmark.

Concrete can be made with high-strength hydraulic lime mortar instead of using cement as a binder. The concrete can be molded in a formwork and mixed with LECA (lightweight expanded clay aggregate) balls for an insulating effect.

In some countries, new binders have appeared on the market that can act as a full or partial replacement for regular Portland cement in the production of concrete and generate a much lower climate impact.

GGBS (ground granulated blast-furnace slag) is a product manufactured from slag, a by-product of iron production. GGBS can replace up to 70% of regular Portland cement when mixing concrete.

Geopolymer cement is another new product that is made from power-plant by-products such as ash or slag; it can completely replace Portland cement.

Insulation materials that can withstand ground contact are often difficult to source locally. Mussel shells are an option if you live near a mussel processing factory. Otherwise, foam glass, a relatively environmentally friendly material made from recycled glass, is often used in ecological construction.

Workshop, where work is exchanged for food and education at the Reforest Farm in Denmark. The straw-bale house is built into a south-facing slope to use the root cellar effect to the north and passive solar heating to the south.

TINA STOUSTRUP JACOBSEN

On the press table at the company Kalle Balle Byg, both thick outer walls and thin partition walls can be made. It simplifies the construction process so that the materials and process for the two types of walls are the same.

TINA STOUSTRUP JACOBSEN

It only took one day to assemble Michael Kallesen's house from straw elements. This included partition walls and openings for doors and windows – ready to be plastered with clay. On the same day, lattice rafters were also installed on the house, so that the felt underlayment could be easily installed. This minimizes the period when straw-bale construction is vulnerable to rain.

How should we build?

Building a house without machines, as was done in the old days, would result in the lowest possible resource consumption and environmental impact. But the use of machines on the construction site only accounts for a small part of the energy consumption in building a house.

The permaculture approach is rather to see the construction of a house as an investment that is made to create a permanent change: a change to a situation where we can live without using fossil fuels in the long run. It is worth investing energy and machine power in this transition when done thoughtfully.

Construction with local natural materials is usually carried out by self-builders or by small craft businesses, and workshops are often arranged in connection with this. The materials can be obtained more cheaply than industrial materials, but on the other hand it takes longer to build with them.

You can get a cheap house by providing the labor yourself, and building your own house is a way to create your own home workplace for a few years. Self-building, including straw-bale houses, is incredibly important for building experience in natural building, but building your own house is a challenging task that requires practical skills and a lot of knowledge in the field. Self-building is therefore not a good solution for everyone.

ELO DAHL MORTENSEN

The School for Life on Møn (an island in Denmark) is built of wood, clay and straw elements from the company Ecococon. The roof is covered with seaweed from the local coast. It is an example of large-scale construction with prefabricated natural materials. Denmark.

Bioregional construction industry

Perhaps the way to a greater spread of natural building with local materials could be the development of a bioregional construction industry, where the bioregion's homes are built from the bioregion's own materials, e.g. manufacturing wall elements made out of local straw and wood. A production plant for straw elements does not have to be particularly high-tech. It can be done under a pavilion, and the construction site will be largely waste-free.

Building with prefabricated wall elements can provide a faster and safer construction process, where we don't have to worry about rain along the way, and it will be cheaper to build a carbon-storing straw-bale house without being a self-builder.

6

Organization

The vision of rebuilding the local communities

The vision is an organization that creates local closed-loop systems, which is a necessity in order to regenerate natural resources, counteract climate change and achieve resilience. It is also the vision that the organization around permaculture projects is designed to last a long time and that it provides well-being for the people who are part of the design: that people have appropriate work and that they experience a form of democracy where they have influence and can help create the framework for their lives.

This section discusses bioregionalism and how bioregions are reestablished, as well as ecovillages and other local communities. It does not address national and international organization, like the UN or the WHO.

The ecovillage Permatopia was ready with 90 homes for move-in in 2018. Here you can see the overall permaculture design for the placement of the communal house, homes, forest, forest garden, meadow, field, electric cars, etc. Denmark.

PERMATOPIA

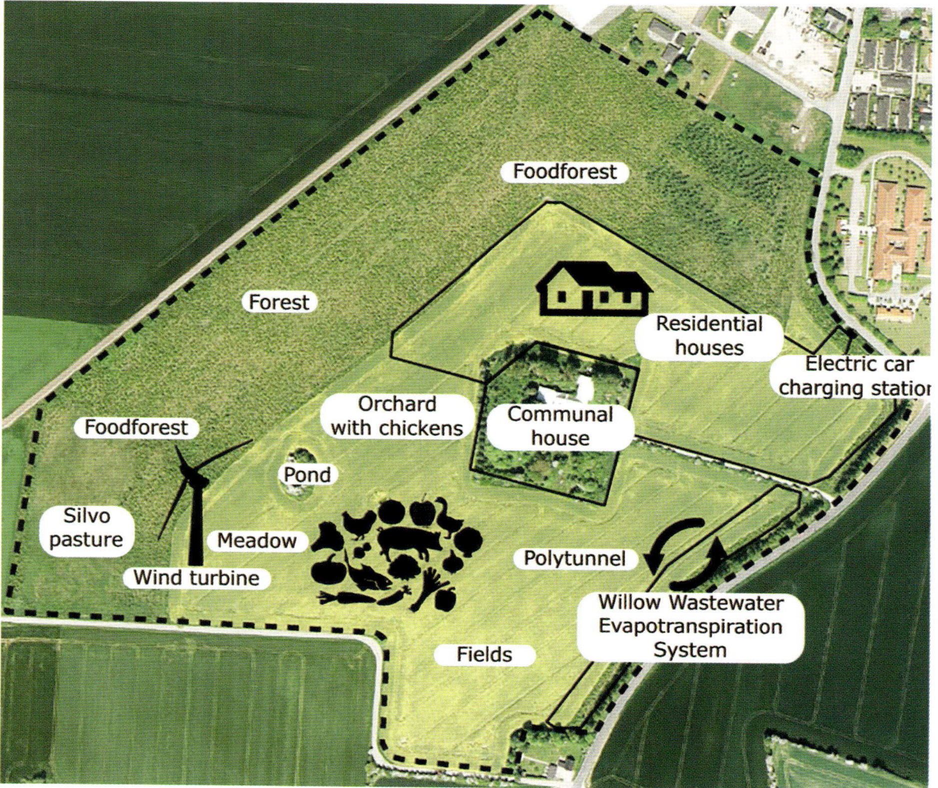

Valnøddelauget (The Walnut Guild) wants to grow, clean and dry walnuts in their local area, as well as make climate accounts and biodiversity measurements. They are established as a non-profit association and this has made it possible to obtain funds for the slow and expensive establishment phase and some of the members have also paid a voluntary 'climate tax' to the association. They currently lease land from two landowners and have planted many hectares. From the beginning, the association has invited the local population, the local agricultural school and associations in the local area to practical work days.

VALNØDDELAUGET JONNA LØVLUND BACH

Bioregion

While we individually develop our permaculture projects and/or work at the political level to deal with the major challenges of regenerating natural resources and counteracting climate change, we must also cooperate at the local and regional community level. This is often underestimated in our current society. Permaculture therefore works with the concept of bioregionalism.

A bioregion is a geographically delimited area, which in some situations is defined by natural boundaries such as watersheds, while in other situations it is more relevant to define it based on the politically delimited areas provided by a municipality or a larger city with its catchment area. The acid test of a bioregion is that people define themselves as being from this area.

According to Bill Mollison, a bioregion can have 7–40,000 inhabitants,[1] but this very much depends on the situation: an island bioregion would be much smaller than an area that includes a city.

The goal is that the cycle of nutrients should be closed within a bioregion and that we can cover our basic needs, such as food, energy, housing and repair of things within it.

In every permaculture project, we must relate to our bioregion and try to ensure that both input and output from our place go to and from the bioregion.

Minimize transport of people

Transport does not have a separate chapter in this book, since the approach to transport is largely to make transport redundant through bioregionalism and strengthening local communities. The places we travel to on a daily basis, such as kindergarten, school and workplace, should ideally not just be in the bioregion but be close enough to be reached by bicycle.

Daily commuting has high costs in terms of time and money, and if we commute with a single person in a car, it can quickly seize a large part of our climate and environmental allowance. For example, on average, a Dane commutes 22 kilometers a day, and this takes up 62% of a person's climate allowance if it is done in a regular gasoline car with one person in it.[2]

If they commute instead in an electric car, the climate footprint will be smaller, but then the consumption of raw materials will instead exceed the personal allowance, due to the consumption of materials for the battery.

Minimize transportation of food

In relation to our climate impact, it is crucial that we eat food that is produced locally as much as possible, as is also suggested in the section on land use for Denmark (see p.124).

The positive climate impacts of locally produced food are illustrated by a study from 2022 based on the concept of food miles.[3] A food mile is defined as 'transporting one tonne of food one kilometer'. The global food system, which consists of food production, food miles and land-use change, accounts for a total of 33% of humanity's climate impact.

The study shows that food miles account for 19% of the food system's

JENS ØSTER MORTENSEN

The Organic Samsø Land Trust works to convert existing farms on the Danish island Samsø to organic. They collect investments as one-off donations, monthly contributions or shares and buy up the farms, which are subsequently leased out to innovative farmers. Here are the tenants of the first two of the four properties purchased, between rows of fruit trees with vegetables underneath, on the farm Søkjærgård that works with permaculture.

climate impact. This is far more than previously assumed. This upward adjustment is attributed to the fact that the study is more comprehensive than previous studies and includes the transport of inputs to agriculture (e.g. chemical fertilizers, machinery and pesticides). Food miles in connection with fruit and vegetables in particular have a high climate impact and account for 36% of the global climate impact from food miles. This is mainly due to the fact that fruit and vegetables must be refrigerated during transport and that they have a high weight, as they contain a lot of water.

The study also shows that although the world's high-income countries only account for 12.5% of the world's population, this part of the world is responsible for a full 46% of the global climate impact from food miles.

We can conclude from the study that if greenhouse gas emissions are to be

ANNA AGGER

Orø Common Land is an association consisting of ordinary citizens and small farmers who operate an agroforestry farm with nut and fruit trees. The association leases and purchases land, and the purpose is to support permaculture projects on the island of Orø. Denmark.

reduced sufficiently, the population in high-income countries must not only move from animal- to plant-based foods, but also start eating locally as much as possible, especially when it comes to fruit and vegetables.

The study also shows that almost all domestic food miles are traveled by truck, which has a much higher climate impact than international food miles, which are primarily covered by cargo ship. Food that is produced in the same country, but far away from the bioregion, also has a high climate footprint for food miles. We must strive for our food to be produced in our own bioregion, and ideally with inputs to agricultural production that also come from the bioregion, so that transport by truck is minimized.

JACOB WESTERGAARD MADSEN

In the nationwide association Andelsgaarde, members pay a monthly amount to purchase farms, which are then leased out to younger farmers who operate various forms of regenerative agriculture. Here you can see an area of agroforestry on Lerbjerggård. Denmark.

Reestablishing the bioregion

In the days before fossil fuels, trade in goods and services was mainly bioregional, but today most things are traded on a global market. When we as permaculturists have to try to cover our needs within the bioregion again, it is obvious to start with our daily service needs, housing and especially food. As explained, switching to carbon-storing permaculture farming is one of the most powerful measures we can take to mitigate climate change. But how do we make the switch happen?

A bioregional guide

A concrete measure we can take to strengthen cooperation and the circulation of goods, etc. in our bioregion is a bioregional guide, which is published every year or regularly updated on a website.

A list is made of which goods and services are available within the bioregion, and where you can buy them. Only products that meet certain criteria and that are locally and sustainably produced are included. Finally, a list is made of everything that is missing in the bioregion so that all needs can be covered. This list of missing measures is important as inspiration for people who want to contribute to bioregional development. It is a concrete way of being 'vision engineers'.

If a local currency is started in the bioregion, which is another concrete tool to support bioregionalism (see the section on monetary systems, p.206), this must also be included in the bioregional guide so that it is clear which places accept the local currency.

Brita Jørgensen and Leif Varmark grow a large proportion of their vegetables, berries and fruits for a plant-based diet in their 400m² garden in Denmark.

Parkens Grøde is an example of how a lawn in a park can be transformed with a permaculture design and then create social community and communication about growing food and niches for biodiversity in the city. Oslo, Norway.

MARCIA KYLE

Transition initiatives

The Transition Network[4] with Transition Towns and Transition Initiatives works specifically with the transformation of entire local communities. In 2006, Transition Town Totnes started as the first initiative, and by 2024, almost 1,000 initiatives are underway worldwide. Vision engineers come together with their positive future visions of how to transform the specific local community within food, energy, construction, transport, education, local money, etc. and the personal inner transformation. Based on these visions, transition plans are made and specific initiatives are launched. In other words, local cooperation stems from how we can reimagine and rebuild our world.

The city in the bioregion

Over half of the world's population lives in cities today. From a positive environmental perspective, residents usually have shorter transport distances, and there is less energy loss from apartments, which have a minimal building envelope, than from detached houses.

The disadvantage is that they constitute a huge zone 0, to which resources in the form of drinking water and fuel, as well as food consisting of nutrients and carbon, must be transported from afar.

In the cities, the resources are turned into sludge, waste and CO_2, which creates pollution and which society has to spend energy and money on handling. And finally, nutrients do not return to the urban catchment area.

In the time before artificial fertilizers were invented, there was often a cycle of nutrients between city and country. This worked by horse-drawn carts or trains driving out of the city at night with the nutrients from the city's privies to the cultivated land that lay around the city, which supplied the city with food. Something similar could be done today under good hygienic conditions.

In the countryside, we can reduce our overall environmental impact further, because here it is easier to achieve self-sufficiency or local and bioregional supply of food, energy and other necessities without transport and with recycling of nutrients and water, and in general, to establish systems that mitigate climate change and rebuild natural resources.

But if we live in the countryside with a city lifestyle, whereby we do not make use of local resources but simply purchase food and energy that has been transported a long way, it will often result in a higher environmental impact than if we lived in the city.

All in all, it is easier to make rural areas sustainable than it is to make big cities sustainable or even resource regenerating.

The vision of the permaculture city

Permaculture's vision is that around every city there should be small, highly productive permaculture farms that supply the cities with food and energy, and that the nutrient cycle should be closed. We must establish rural–urban links.

Finally, in cities, intensive cultivation can also be established in park and garden areas, allotments and private gardens for self-sufficiency using the cultivation techniques described in this book.

In the mid-1800s, Paris was self-sufficient in vegetables within the city limits,[5] and in Havana, Cuba, in the 1990s, 80% of the vegetables were produced in such city gardens.[6] Of course, how much of the food can be grown depends on how densely populated the city is, but there is undoubtedly a large untapped potential.

The economics of agriculture

Land and property ownership

The largest 1% of farms operate more than 70% of the world's farmland.[7] Today, farmland around the world is being bought up by corporations and private equity funds as an investment. In some cases, it is not possible to find out who owns the land in our local communities. Only a small portion of the world's farmland is owned by families and operated as small farms.

The consolidation of land into a few large farms, as well as the enormous increase in the value of farmland, makes it difficult for younger people to acquire farms and properties. This is further reinforced by the fact that older generations own the vast majority of the capital in society. These conditions deprive local populations, and especially younger people, of the opportunity to influence local development.

In order for us to transform our local communities and our agriculture and

ØSTERGRO

Østergro in the middle of Copenhagen has raised beds, bees and a restaurant on the roof of a high-rise building.

Once agroforestry systems are established, they can be very permanent. In this oasis village in Morocco, small fields can be seen between 2000-year-old agroforestry systems of dates, olives, pomegranates, almonds, etc. In this particular village, they have decided not to sell to Europeans, who buy property in the nearby villages for exotic winter summer houses. They decide things at village meetings, and the land is inherited. The products are for self-sufficiency and direct sale at the market in the larger port town nearby.

restore the circuits to the cities, it is necessary to own land and properties. It is a prerequisite for us to have power over the direction in which we develop society, and thus to create permaculture.

We can establish countermeasures by joining forces for joint ownership, creating foundations and associations that buy up land and lease it out to farmers, and we can do something similar with production facilities, such as setting up jointly owned greenhouses, nurseries, sawmills, etc.

The farmer's conditions

The farmer is central to the transition to permaculture, but the farmer is often bound by a large debt and a low salary.

The proportion of the money consumers spend on food that ends up with farmers has fallen significantly since the 1950s.[8] When apples and carrots are sold in the supermarket today, 36% and 26% of the sales price respectively ends up with the farmer, the rest goes to the supermarket and the wholesaler. When a loaf of bread is sold in the supermarket, only 8% of the sale price ends up with the farmer who produced the grain; the rest goes to the supermarket, bakery and mill.[9] After farmers have paid their costs in connection with production, the profit is therefore small.

Being a farmer today is often a low-paid job with long and often irregular working hours and with a high risk in the form of weather fluctuations and fluctuations in prices on the global food market. The farmer is in direct competition with, for example, underpaid farmers in distant countries or large companies that use child labor and illegal immigrants who work in slave-like conditions.

The current system does not care for the farmer. And the structure in which the farmer is entangled does not allow for a change to carbon-storing and highly productive cultivation methods. On the contrary, the farmer will often be penalized financially for initiatives that work towards permaculture farming.

The food system's externalities

The underlying problem is that the environmental, health and social costs are not included in the price of the food we can buy in stores; these costs are called 'externalized'.

A research group has worked to calculate the true price of food in the global food system.[10] They have included a number of environmental and health externalities in the price, including greenhouse gas emissions, land use, freshwater consumption, air pollution and contributions to cardiovascular diseases, and have concluded that they result in costs of 19.8 trillion USD annually.

If we compare this with the fact that food is sold for nine trillion USD annually, we can see that food prices would have to be about three times higher on average if we had to pay the true price for the products produced by the existing food system.

The members of the Copenhagen food co-op take turns packing bags of seasonal vegetables. The co-op was started by consumers who want a different kind of agriculture and fair trade for the farmer. Denmark.

ANGEL QUINTANA

This calculation even omits important externalities, such as the social costs of underpaid and child labor, soil degradation, land use for grazing, other forms of air pollution than NH_3, antibiotic resistance and zoonoses. Externalities related to food transport and processing are also not included. Externalizing the adverse side effects of the global food system creates a systemic undercurrent in the economy that continuously opposes the development of agriculture that cares for the earth and people. The economic framework gives an enormous competitive advantage to the farms that pollute the most and produce unhealthy food because they do not have to pay for the negative consequences of their production. For permaculture farming, which mitigates climate change and rebuilds natural resources, the competitive conditions are particularly bad because the positive externalities, such as carbon storage, are also not included in the economy.

Proposals to convert agriculture to permaculture are often met with the argument that permaculture farming is not a sufficiently efficient way of producing food, and that there are few examples of economically viable permaculture farms. One response to this criticism is that if we convert the economy to align it with the true price of food, then industrial agricultural products would become three times as expensive as they are today, or more, and products from permaculture farming would probably cost the same as they do now. In other words, it would completely turn what was competitive upside down.

It is in the interest of society as a whole to convert the economy to take into account the externalities of agriculture, not just in the long term, where we would avoid the environmental consequences of current production, but probably also in the short term, since a large part of the calculated externalities are about health. The research group behind the study writes that with true pricing, significant savings in public spending can likely be achieved through lower health costs, avoided environmental mitigation measures (such as climate change) and reduction of subsidies. These savings could be enough to make food cheaper than it is now, even after environmental and social costs are internalized.[11]

Agricultural subsidies

An obvious place to start is to adapt the schemes for agricultural subsidies that already exist so that they reward climate- and environment-friendly farming. A 2021 World Bank study has shown that only 9% of agricultural subsidies paid supports environmental protection.[12]

In fact, many permaculture farms fall outside the grant aid rules altogether and do not receive the support allocated to industrial monoculture farming at all. It is also unfortunate that the existing support rules in the EU largely steer farmers away from making the best possible permaculture design but instead turn the design towards a simplified version with fewer species that does not take into account the microclimate and landscape profile. This means that several permaculture principles are not followed, such as the planning tools, diversity, beneficial relationships, multiple dimensions, etc. It is unfortunate if we as permaculturists allow such subsidy schemes to guide our designs, because permaculture and agroforestry systems are designed to function for hundreds, perhaps even thousands of years. Grants should instead be designed to support good permaculture design.

RUTH MARIE KONDRUP

At Kastanjely, they grow a no-dig market garden and have recently started establishing agroforestry. They sell both from the farm shop and through a subscription scheme. Customers are regularly invited to tours and communal meals. Denmark.

Consumer responsibility and influence

As permaculturists who buy our food, we must try to navigate an economic system that, from a permaculture perspective, creates a completely flawed pricing system. Consumers who have a financial surplus need to actively support the transformation of agriculture, to take their share of responsibility, even if in some cases it means that we have to pay a slightly higher price for our food.

Today, it is difficult as a consumer to obtain food that has a low ecological footprint and mitigates climate change, and it is important to help those permaculture farms that are starting up to survive in the existing market. These are pioneering projects that we will all benefit from if they are successful and are replicated.

Rural-urban links

Direct sales from the farmer to the consumer, where costly intermediaries are avoided, are often crucial for small permaculture farms to be economically profitable, and at the same time it can help to ensure that the price difference for the consumer is not too large.

By establishing rural-urban links with direct sales of agricultural products to consumers, it is also possible to minimize transport, and the consumer can learn about where and how the product was produced.

Direct sales are easiest to establish if the farm is located in reasonable proximity to a larger town where there are consumers who are interested in buying the products.

With rural-urban links, the consumer gets the opportunity to take responsibility and help determine the development of agriculture. In some cases, it can be a form of fair trade – but in rich countries.

MOSEGÅRDEN, TANNIE NYBOE

Mosegården's community-based farming consists of those who run the farm and all those who have bought a share of the year's harvest. The aim is to create a stronger community around the production and marketing of seasonal vegetables, where the cooperative members get a share of the year's harvest with a weekly bag, and the farmers get greater security and financial stability in production. The farm is located 65km from Copenhagen, and weekly veg harvests are delivered to pick up points in the city. Denmark.

Concrete ideas for starting direct sales of agricultural products

First, we describe the forms of organization that most resemble today's common practice, and then come solutions where consumers take increasingly greater responsibility.

Farmers market – Farmers go to the city and sell their goods themselves. This provides good direct contact with the customer but is often very time-consuming for the farmer, who also has no guarantee that the goods will be sold, since the consumer is not obliged to buy them.

Farm shop – Selling directly from a farm or a large garden can be an option if the farm is centrally located, both so that there is enough turnover and so that it does not result in extra car driving for the customers. Here, the consumer gets the opportunity to see the farm on which the goods are produced. The farmer still risks the goods they have grown and harvested not being sold to the uncommitted consumer.

Food co-ops – In food co-ops, consumers come together and buy goods directly from farms within a local radius of the city. Consumers receive the season's production and participate in a rotation scheme to pack a weekly box for all members of the food co-op. This provides better economics for the farmer and for the members of the food co-op. Everything the farmer has harvested is sold, as agreements have been made in advance. It is worth considering whether to hire someone to pack the produce, as experience shows that although they would like to, busy families with children do not have the resources to participate in running a food co-op.

Box sales – The consumer is delivered a box of seasonal fruit and vegetables from a farm. The farmer knows in advance what is to be harvested, and if a subscription has been signed, the farmer is assured of his sales. The consumer saves time on shopping. Here, part of the sales price goes toward the delivery and packing of boxes.

Subscription farming – Consumers participate here and provide the farmer with even more security by signing a one-year subscription to the farm's production in advance. Consumers thus share the risk of both good and bad growing seasons and in this way create safe working conditions for the farmer.

Cooperative farming – Here, consumers take full responsibility and become co-owners of the farm, often via a trust that owns it. In this way, consumers can have influence and help determine what and how to grow and produce. Consumers can help plan the year's production and hire a farmer who takes care of the operation for a fixed salary. Often, the employed workforce is combined with shared working days, which provide social interaction and the joy of taking part in rural life. For many farmers, designing and planning the operation of the farm is a great satisfaction and joy, and we should be aware, in decision-making, that the farmer usually has greater professional knowledge and experience than the consumers. In some cooperative farms, the farmer is therefore given a high degree of autonomy over the operation of the farm, but a more climate- and environmentally friendly operation is enabled by renting the farm to the farmer on favorable terms.

Subscription farming and cooperative farming are two versions of what is also called Community Supported Agriculture (CSA), where the gap between consumers and producers is reduced and instead the risk, responsibility and rewards associated with farming are shared. Between 500,000 and one million people in Europe today get their food from CSA farms.[13]

Ecovillages and local communities

Many who want to work with permaculture choose to live in local communities of various forms like cohousing and collectives, commonly referred to as ecovillages. For this reason, we have included this section on the organization of ecovillages.

The advantage of an ecovillage is that we have someone with whom we can share our belief to create a different culture with different norms based on the permaculture ethic. In addition, it is also practical to be able to work together and share things with each other, such as having a car-sharing scheme, making communal meals or growing food together.

Ecovillages are often, but not always, based on permaculture, and the following applies to ecovillages in general. When we commit ourselves to each other and are going to live together, work together and generally run things together, it is important that we design good organizational structures, just as we must create a good permaculture design for the physical landscapes.

It is important to emphasize that we do not necessarily have to live in an ecovillage to work with permaculture; it is up to one's personal preferences. If we choose to live as a family or alone, we can, if we wish, create other local communities such as transition towns, permaculture study groups, community gardens, farming cooperatives, cooperative stores, etc. Such projects can be organized with more or less close cooperation.

Diana Leafe Christian, who lives in the ecovillage Earth Haven in the USA, has written the book *Creating a Life Together*[14] based on interviews with a large number of ecovillage initiators. What follows is mainly inspired by this book.

In the book, she describes a number of conditions that must be in place in order for one to be successful in establishing a new ecovillage. These are conditions where, if they are not in place, they can function as 'time bombs' that can explode later and result in structural conflicts that have to do with the way we are organized, as opposed to personal conflicts.

How do we avoid structural conflicts?

Based on Diana's book, the following will review the five most important aspects of how we can avoid structural conflicts in ecovillages. Some of the aspects will also apply in connection with the start-up and organization of other kinds of local communities.

Find the common vision and create vision documents

Just as when we design for individuals or a family, we must also clarify for communities what the vision of the project is. There is probably no more destructive reason for structural conflict in a community than when it turns out that the members have fundamentally incompatible visions. This can result in many different conflicts, for example, about how much money or how much work we should spend on different projects. We can never make a decision that everyone is satisfied with, because we all want to go in different directions.

Everyone must know the common vision and support it. The common vision should be discussed, decided and written down from the outset. The vision documents can usefully contain a short vision statement that extracts the essence of what the project is about and a longer version of what we are aiming for together. In addition, there can be one or more documents on several pages that describe goals and strategies for the project. It is the vision that should hold the community together when it encounters challenges along the way.

Another option is to organize the project so that the members are obligated to some conditions imposed by an external body, e.g. an external foundation board, public authorities or similar. This can make the shared vision extra strong and able to withstand pressure from the surrounding culture. In the world's high-income countries, the ethic of fair share is particularly relevant. Surrounding society sets norms about overconsumption and the waste of resources, so it can be difficult to stand firm on a vision of reaching a sustainable level of consumption.

Make decisions in writing

In addition to creating vision documents, it is generally important to have decisions in writing, and the legal aspects must be clearly described with a set of statutes that explain how we handle moving in and out, exclusion, making decisions, etc.

Find a democratic and involving decision-making process

It is important that everyone can understand the decision-making process. Overall, we must distinguish between direct democracy, where everyone in principle helps make all decisions, and representative democracy, where someone is elected to make decisions on behalf of the community.

One criterion for good democracy is that there is always a debate before decisions are made. This applies to both direct and representative democracy. In larger contexts, where many and complicated decisions have to be made, it is usually best that we use representative democracy. This is simply because direct democracy requires too much time and energy, and representative democracy gives us a necessary division of labor. If we try to exercise direct democracy in a large group, the consequence may, for example, be a distortion, where those who have the least time due to work, illness, parenthood and the like have less influence than those who have a surplus of time.[15] That said, direct democracy is a good solution in smaller associations, such as ecovillages.

Consensus democracy – Many ecovillages use the principle of consensus democracy, a principle mentioned by the philosopher Thomas Hobbes as far back as the 17th century. Consensus means that decisions are made by talking our way to solutions that everyone finds acceptable. This should not be confused with everyone agreeing and thinking that the decisions made are always the best.

The advantages of consensus are that it creates dialogue and promotes cooperation compared to majority voting, where you can in principle skip all discussions and go straight to a vote, after which a majority of 51% can choose to ignore a minority of 49%. Such a vote may be faster in the first instance, but subsequently it will often mean that the minority that has been ignored will be slow to implement the decision or will directly oppose it. It is through dialogue that we come to understand each other so that we can show consideration, make compromises and work towards the same goal.

When we work with consensus democracy, however, it is particularly important that we have a common vision, as otherwise it can be very difficult to make decisions. We must also be careful that consensus democracy does not turn into what Diana Leafe Christian calls 'pseudo consensus' where one or more of the participants has an 'I know best attitude' and lacks the flexibility and ability to adapt to others.

It is important that we do not misunderstand the right of veto that one has in a consensus democracy. Exercising a veto in a consensus democracy is something that one only does a few times over a lifetime, and only if one finds that a proposal is unethical or unsafe. If you repeatedly feel the need to veto decisions, it may be a sign that you need to consider whether you share the common vision or whether perhaps you are part of the wrong community.

Another danger of consensus democracy is that it becomes too time-consuming so that only some people have time to participate. Decisions should not be the result of a struggle for endurance,

In Svanholm Storkollektiv (Svanholm collective), 4,158 decisions have been made over 45 years using consensus democracy alone. Democracy has developed so that today, for example, there are debate meetings, community meetings and several types of subgroups, as well as a special group that plans the community meetings.

SIV WERNER

such as all-night meetings. It is often a good idea to send decisions to committees so that only those who have strong opinions on an issue or are particularly affected by it have time to consider the issue before the decision is made collectively.

Consensus is not always the best form of decision-making, especially in loosely knit communities. But the alternative does not have to be majority voting. We can describe in the statutes procedures for first trying to reach consensus decisions at one or more meetings, after which there is the possibility of sending the decision to a vote, and we can stipulate that the decision is only made if, for example, 75% of the members or all but two vote in favor. In this way, we ensure that the broad majority supports the decision and that the community is not split down the middle.

Sociocracy as a form of government – The consensus-based democracy has been further developed into the framework system sociocracy, which sets up an entire organizational structure. The first modern implementation of sociocracy was developed by Gerard Endenburg in the late 1990s as a new method of managing companies, but the method has since been used in public, private and nonprofit organizations, as well as in professional contexts.[16]

First, it is important that the organization has a clear purpose/vision as described above. But what is special about sociocracy is that it sets up an entire structure consisting of well-defined circles, which is the sociocratic name for a group. There is a common circle in the center of the organization, and from here there are subcircles, which in turn can have more subcircles. Circles can be created and closed continuously as needed. Each subcircle has clear objectives, domains with responsibility and a mandate to act within their subarea. In this way, each circle is able to make decisions autonomously within its domain so that decisions are made quickly and thus create efficiency. A circle typically has four to eight members.

Within the circles there are people with well-defined roles:

- A leader, who is responsible for operations and ensuring the purpose
- A facilitator
- A secretary, who is responsible for minutes and other documents
- A delegate, who communicates to the circle above.

The leader and delegate are 'linking roles', who are also members of the circle above; this ensures information passes between the circles. The common circle consists only of linking roles.

Sociocracy is a form of government that uses a consent-based decision-making process among individuals, where everyone in a circle must think the proposal is 'good enough to be tried', and this should not be confused with it necessarily being one's preference. Sociocracy also sets up structures for evaluating meetings, roles, decisions made and purposes. This ensures that we continuously adjust and improve the organization based on real experiences and makes it easier to consent because we know that an evaluation will be made.[17]

Learn good communication and group-process skills

Good communication skills are characterized by our ability to talk about sensitive topics and still feel connected. It's about practicing listening to what others say, even if you disagree with them or if you are criticized, and constantly thinking about whether there is a kernel of truth in what the other person is saying. It's also about being careful about how we give criticism so that we stay on our own side of the fence and describe how we ourselves experience something someone else has done, rather than attributing bad qualities to the other. Finally, it's also about daring to hold each other accountable for making sure that joint decisions are actually followed through.

Be selective when accepting new members

When we move into a community together, we become obligated to each other and have to make many things work together, hopefully for many years. These are relationships that have emotional aspects, just like in a marriage, and at the same time have practical and

Examples of vision documents

In the ecovillage Friland, they have formulated some principles very briefly, which all members know. The key points in the paper are based on

- building your own house without debt
- starting an independent business
- purifying your own wastewater
- not receiving subsidies
- and otherwise reducing your energy and resource consumption as much as possible.

The permaculture farms in Skovvirke have been granted a rural zone permit on special terms from the municipality, which, among other things, ensures that group members will use permaculture farming and lead a lifestyle with a relatively low ecological footprint. The terms cannot be changed by anyone in the group.

Terms for using the permit:

1. Farming must be carried out with production for sale – as well as other production that uses the site's resources. Those who live on the farms must cover 75% of their household needs through the use of the site's resources.
2. Those who live on the farms must live a lifestyle with an ecological footprint that does not occupy more than 2.4 global hectares.
3. The site must be operated fossil-free. (Fossil fuels may be used for the construction of the site, but not for operation.)
4. The carbon content of the soil in the agroforestry must be measured. Soil tillage may not be carried out on more than 10% of the area.
5. The conditions for biodiversity must be promoted.
6. Buildings must be built so that they blend into the landscape. The buildings must be built from local natural materials. The living area per farm must be a maximum of $80m^2$ internal dimensions. A compost toilet must be included.
7. The homes must be built as applied for.
8. The homes cannot be subdivided.
9. After no later than five years, the homes must be hidden by vegetation in the summer, when there are leaves on the trees.

organizational aspects because together we have to solve a wide range of work-related tasks, just like in a company.

Therefore, it is important to select who we let into the community. As described above, it is important that the project has a common vision, and we must therefore, among other things, select whether new applicants support the common visions.

In addition, it is also important that we only admit new members who have a basic emotional maturity that makes them able to live in the community. This does not mean that we should expect everyone to be a perfect person – most people have ups and downs in life, but severely unmet emotional needs, possibly in combination with substance or alcohol abuse, in even just one person in a group can throw the entire community out of balance, especially when it comes to smaller or newly started ecovillages. The person's emotional pain can play out in community meetings on completely different topics and generally be expressed in behavior that burdens the rest of the group.

Conflicts can also arise between others in the group about how the person's behavior should be handled. This has often caused otherwise satisfied members to leave ecovillages and can even cause communities to disintegrate. It can also end with the person being excluded, meaning they have suffered yet another defeat that it might have been better to avoid.

But how can you tell whether an applicant has severely unmet emotional needs? You probably get the surest signs by looking at the applicant's past. If there has been a long series of conflicts and breakups, it is quite likely that they have and that moving into a new community will not immediately solve the person's problems. It is also important that we create a well-designed admission procedure, where it takes some time to get into the group so that we get to know the applicants well; we can possibly use a well-thought-out questionnaire that is completed during the application period.

Some may object that communities should be open and inclusive, and that this is how we live out the ethic of people care. That is also true. But if we take someone into the group with severe unmet emotional needs, we must be aware of it. It is much easier to deal with recognized psychological problems than unrecognized ones.

It is entirely in line with the permaculture ethic to define an ecovillage as a therapeutic project, but we must then be aware that this is what we are doing. We must ensure that we have the necessary resources for the task and be aware that the community will then have fewer resources for some of the other aspects of permaculture.

It is not an easy task to organize a new ecovillage based on permaculture, but it is both an exciting and socially important task. As idealistic and eager permaculturists, we may end up focusing only on the physical changes, such as planting trees to counteract climate change. But we must also spend time creating a solid organization; there are unfortunately many examples of permaculture projects only existing for a short time, which contradicts the long-term regeneration that is precisely the purpose of permaculture.

7 Monetary Systems

The vision for money that promotes permaculture

The subject of economics stands out in permaculture by not directly dealing with the regeneration of natural resources and the mitigation of climate change. However, the economic structures that exist in a given society have major consequences for our options for action and change: the economic system promotes and counteracts different actions. In other words, we cannot ignore our economic system, which often systematically contradicts the changes we are working for in the permaculture movement. Therefore, there are several battles that need to be fought in this area.

The vision is to create a monetary system that promotes local trade in sustainable goods without overconsumption and degradation of natural resources, thereby helping to counteract climate change. The monetary system must also encourage us to help each other and thus achieve a strengthened local community. The monetary system must support the ethics of permaculture and contribute to caring for the earth and people and creating fair distribution.

Doughnut economics

Economist Kate Raworth has created the 'Doughnut model', which is a vision for a new way of thinking and calculating economics. The model is becoming increasingly popular. It is similar to the

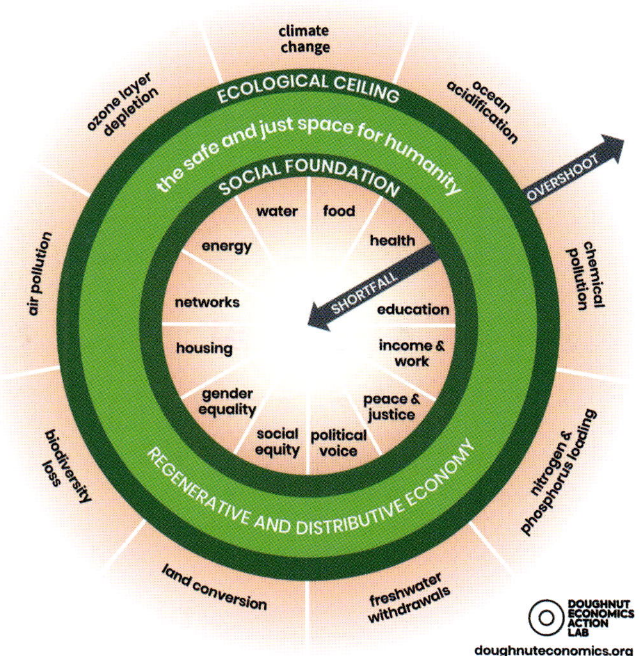

KATE RAWORTH

A modest self-built house made of natural and recycled materials has enabled Morten Hylleberg and Hanne Lerstrup Pedersen to establish themselves debt-free in the housing market. This, combined with the desire to have low consumption, has meant that they can manage on one part-time income and therefore have time for much else. The family lives in the ecovillage Fri og fro in Denmark, where debt-free and low-consumption living are part of the vision.

EISTRUP FILM

permaculture vision, although slightly different concepts are used. The doughnut, with its outer circle, illustrates care for the earth in the form of eight planetary boundaries that we must not exceed. The inner ring illustrates care for people in the form of the social foundation where people have their needs met. The doughnut-shaped area between the two rings is the ecologically safe and socially-just area where people can thrive.

Structure of this chapter

In the following, we will first delve a little deeper into the current monetary system and the problems that are built into it. The monetary system is politically determined, and therefore we as permaculturists can engage in politics to change it, and here we will present various organizations that work toward alternative models. We mention a number of concrete ways in which we can organize ourselves and practical measures we can take to change the situation, as well as how we work in practice within the current monetary system.

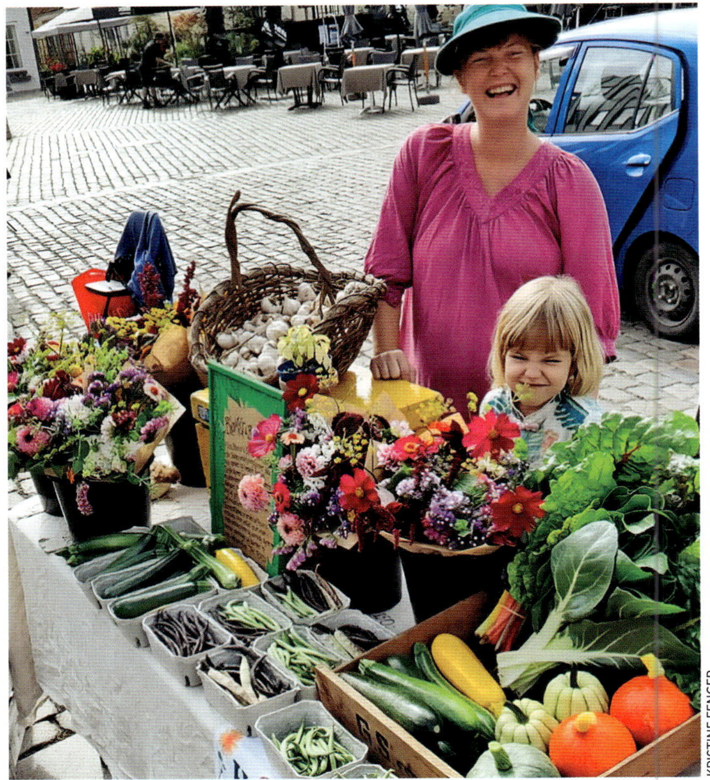

KRISTINE FENGER

If we want to live from small permaculture businesses with a great diversity of income sources, this will, for most people, be linked to a low-consumption lifestyle, which is in line with a low ecological footprint.

Kristine Fenger and her family have bought a small country house and started a small company selling vegetables, nursery plants, flowers, herbal tea, garden visits, etc. Here she sells at the market in Fåborg, Denmark.

Modern money

What is money and its function?

Throughout time, tribal people have bartered between two parties without money. In contrast to the barter economy, money is a unit of measurement that enables goods or services to be transferred between several parties.

The basis for something to be defined as money is that it is accepted as a general means of payment in the economy.

Throughout human history, monetary systems have been designed in all sorts of ways. Different monetary systems promote or discourage different actions, products, emotions and values.

Modern money is created by banks[1,2]

Today, there is no money that banks have not created through lending, and the issuance of money and banknotes through the central banks only accounts for a few percent. Both printed money and account money are created when a customer takes out a so-called 'loan'

POSITIVE MONEY

Does the bank lend out money that already exists? Or does it create new money?

from the bank. But the borrower does not borrow anything, because the money is created out of nothing and disappears again when the debt is repaid. Banks thus create money when loans are issued.

Virtually all money circulating among us has thus arisen because a citizen has incurred debt to a bank, which must be repaid with interest. Interest is a profit for the one who has the right or power to create money, even though it is not a big job to enter a few digits into an account.

A few decades ago, it was the central banks and therefore the states that alone had the right to print banknotes and make coins. But the account money, which today constitutes over 95% of the money supply, is printed by private banks. This means that the profit from interest has gone from accruing to the state and society to today accruing to the banks and their owners.

Money with interest creates inequality

Interest creates an inevitable process in which money is transferred from all those who have debt to those who have wealth. In practice, a smaller part of the population has greater wealth, and the rest owes money.[3,4]

Money is becoming concentrated in the richest part of the population without them working for it. In 2014, the Danish population paid 200 billion DKK directly to the banks in interest and fees. In comparison, the healthcare system costs 150 billion DKK and the education sector 125 billion DKK. 100 billion DKK of the 200 billion DKK was paid out as interest to the banks' richest customers, 50 billion DKK to the banks' employees and 30 billion DKK to the banks' shareholders.[5]

The population of the United Kingdom currently spends an average of around 4% of their earnings on paying interest on household debt.[6]

We also indirectly contribute to paying for agricultural loans when we buy food; landlord loans when we pay rent, etc. because interest helps to increase the general price level in society.

But if we belong to the privileged with a surplus of capital, we can help break the circle by putting our money in fund-owned permaculture projects that are not designed to provide a return to investors, for example, a land trust that buys up land for permaculture farming.

The concentration of money among the richest is further reinforced by speculation in currency, stocks, the housing market, etc., and this speculation today accounts for over 95% of currency transactions.[7] Likewise, the process of concentration of money today also goes

Interest is a systematic transfer of money from the many who owe money to the few who have money in the bank. This is further reinforced through trading in shares and currency speculation, etc.[8]

from poor countries with debt to rich countries as a systematic maintenance of inequality.

Economist Thomas Piketty has shown how, on a global scale, in recent decades it has become increasingly difficult to achieve wealth based on work and increasingly easy to achieve wealth based on capital, and that the more capital you possess, the more you earn per dollar or Euro you invest.

This means that the monetary system we use today is in direct conflict with the permaculture ethic of fair share. For most of us, the current monetary system leads to scarcity (debt) and negative stressed feelings and thoughts about how we will get money to pay our debt and interest expenses.

For many permaculturists, the monetary system is also a barrier to getting started, as it is expensive, for example, to start a business or acquire agricultural land. Over 75% of the money that banks create goes to trade in housing and agricultural land, which contributes to pushing up land prices, especially when banks constantly create more money than is actually needed. Interest can be seen as a tax paid to the banks on the money they create out of nothing.

The welfare state with taxes and social benefits counteracts the concentration of funds as a symptomatic solution, but it does not change the monetary system.

Today's monetary system is dependent on eternal growth[9,10]

As mentioned, there is no money that the banks have not created. The borrower must repay their loan with interest, but the extra interest money has never been created. The borrowers must therefore compete with each other to try to obtain the extra money. It is only possible for all borrowers to obtain enough money if there is growth, and more and more money is created by more and more loans. Otherwise, someone must go bankrupt so that others can repay their debt and interest.

Growth is therefore built into the central monetary system as a necessity. This is further reinforced when shareholders must make money from the increase in value of their shares in companies – companies must achieve growth. But when growth increases, the consumption of resources and the emission of CO_2 increase. If we try to reduce consumption in order to limit environmental and climate problems, society will experience negative growth, and someone will have to go bankrupt.

But infinite growth in the consumption of resources is an impossibility within the Earth's finite ecosystem. The monetary system has thus helped to promote the overconsumption of the Earth's resources, and we see a contradictory relationship where environmental and climate problems cannot be solved without creating a crisis in the global economy. Using the current monetary system and solving environmental and climate problems are mutually contradictory.

A new economy

The following section will present different proposals for how we can create a new economy, at the individual, local and national level.

Positive money

The monetary system can and should be changed, and a large part of the problems lie at the national and international level. The association Positive Money[11] exists in many countries and works for three solutions to change the monetary system:

- Money issued by an independent democratic body
- Money created by national banks without debt and thus without interest
- Money that goes into the real economy and not into housing bubbles and speculation.

Positive Money is thus an organization that works for the same goals as permaculture, but at a political level.

Local complementary currencies

Creating local complementary money as a supplement to the central money is a tool often used in connection with permaculture. The local money is designed so that it functions within, for example, an ecovillage, a city or a bioregion. Locally owned stores are often willing to support the local economy and accept payment in local currency. Chain stores, where almost all the goods they sell come from other parts of the world, on the other hand, will usually not accept it, because they cannot use local money to purchase their goods.

If local money can only be used in local stores, it will help promote local circulation of goods with limited transportation, rather than international import and export to multinational corporations – which in turn concentrate wealth in the hands of the richest.

Local circulation of goods creates a stronger local identity and sense of community and wellbeing because we can see where the goods come from or who they are sold to, and because we provide services for neighbors we would otherwise have no relationship with. Local money creates more resilience to economic crises because it gives people who do not have access to money from the central monetary system the opportunity to sell, buy and provide services in the local community. This gives a local community the power to take action and set new priorities, and the opportunity to give local communities greater influence. For these reasons, local money systems also tend to emerge in times of crisis. Local money has existed since the crisis in the 1930s, when many local areas in countries such as the USA, Germany and Austria created their own money.

LETS (Local Exchange and Trading System) is the simplest model. Here, an account is created where everyone is basically at zero. When trading, the person who receives a service or goods will then 'commit' to, for example, three units of the local money, while the seller of the goods receives three units of the local money. The word debt is not used, as the system depends on the fact that as many people have 'committed' as those who have received money, and the total sum will always be zero. The currency arises in reciprocity when it is used, and is interest-free, which means that the person who has committed gets an interest-free loan. The system can be used to trade many different things, from jam to childcare to handicrafts, etc.[12]

Activity, wellbeing and resilience

In a situation where the inflow of money to the local community becomes too small, so that we do not have money to trade with internally, or where there is no trust in the national currency due to rapid inflation, a local currency will make a big difference because it can be kept going, albeit at a reduced rate.

Experiments with local currencies show that they often have greater circulation and thus create more activity than national currencies. At the same time, areas with local currencies can partly decouple themselves from national and global financial speculation. This makes regions economically more resilient to crises that are triggered in distant places but still harm the local population.

Some local currencies have a bio-regional directory or a newspaper where users advertise their needs and services to everyone, and others are systems where trading is simply done hour by hour, and the value of goods can be converted into time. Certain local currencies work with printed banknotes, some are digital and others both. The important thing is that it enables trade between more parties than when two people exchange something.

Digital money is an easy way to trade, which can help increase the use of the local currency, but we must be aware that digital currencies are less resilient because in the event of a power failure, it will become impossible to trade from one moment to the next.

Examples of a local currencies

There are many different initiatives with local currencies around the world. Some that work and others that have had challenges and had to stop again. Here are three examples.

WIR – WIR is a currency and exchange network that operates throughout Switzerland. WIR is an abbreviation for Wirtschaftsring, which means Economic Circle.

WIR emerged as a reaction to the economic crisis of the 1930s and is the world's oldest local currency. It aims to support small and medium-sized businesses through a credit system that supplements the official currency.

The purpose of the WIR system, still today, is to strengthen liquidity and economic stability among participants – especially in times of crisis when cash is scarce. That is why the system also offers WIR credit, which functions as a form of interest-free or low-interest loans in WIR currency.

The currency in the WIR system is electronic, cannot be exchanged for Swiss francs and is only used internally in the network. Members of the network can trade with each other by paying in whole or in part in WIR, which can ease the pressure on cash and encourage local trade.

Participants in the WIR system are typically small and medium-sized enterprises, including sole proprietorships such as craftsmen. WIR is used by approximately 60,000 companies and is therefore actually used in the economy.

Chiemgauer – Chiemgauer is a regional, complementary currency in Bavaria, Germany, which started back in 2003 and aims to strengthen the local economy,

CHIEMGAUER E.V

Chiemgauer is a regional currency. The motif of the banknotes shows insects from the region and bears the text 'For a new coexistence'. Chiemgauer uses demurrage, a fee that encourages people to spend the money locally and quickly, instead of hoarding it.

It works like this: the owner of a banknote must buy a small sticker showing that the fee has been paid and stick it on the fields at the bottom of the banknotes in order for them to retain their full value.

support sustainability and promote civil society.

Chiemgauer is available in both physical and digital form. It is put into circulation when a user exchanges euros for Chiemgauer at a ratio of 1:1. The user can then trade with local businesses that accept Chiemgauer as a means of payment. Businesses can exchange Chiemgauer back into euros for a fee of 5%, which motivates them to continue using the network rather than converting. This strengthens local trade, and it is typically environmentally conscious citizens, local producers and small businesses that use the currency. 3% of the amount goes to a charitable organization of the user's choice (e.g. schools, sports clubs and environmental groups). To keep the money in circulation, it has a 'demurrage', a kind of small negative interest rate that encourages quick use.

Since 2010, it has also been possible to get small loans at cheap interest rates in Chiemgauer. In 2022, there were almost 4,000 individuals and more than 400 businesses using Chiemgauer.

Bristol Pound – In 2007, the Totnes Pound, a local printed banknote, was launched as part of Transition Town Totnes and inspired several local currencies, especially in the UK. The Bristol Pound became the largest, and in 2016 had over 2,000 individual users and businesses and could be used for train and bus journeys, and to pay tradespeople for work carried out, council tax and even employee wages. The Bristol Pound was from the start both a printed and digital currency. With over 1 million Bristol pounds in circulation, Bristol's economy became the fastest growing outside London in the UK.

But even with this successful local currency, there was a tendency for it to be used mostly in subcultures and less by mainstream tradespeople etc. Many notes were also bought as souvenirs by tourists.

Interest in and use of the Bristol Pound declined after 2016, perhaps due to the inconvenience of having multiple types of banknotes. The Bristol Pound finally ceased in 2020 when Covid reduced local trade.

Summary – Like the Bristol Pound, there are several other examples of local currencies that fail or are only used very little, despite a lot of effort being put into starting them. This does not mean that we should not try but that we know the challenges and strive to reach out more widely with local currencies so that we achieve a community of users with a wide variety of goods and can effectively reduce our dependence on national currencies.

Carbon trading

Local money can help rebuild local communities and bioregions, but it does not provide a systematic solution to the problem of externalities such as biodiversity, clean groundwater, etc. in the economy, which was discussed in Chapter 6. Carbon trading is a way to get some of these externalities included in the economy.

In the European Union, the United Kingdom and a number of other countries, power plants, aviation, shipping and the most emitting industries must pay for their greenhouse gas emissions by purchasing CO_2 quotas.

In addition, there is a voluntary market for CO_2 credits. Here, companies can pay for efforts to counteract climate change, such as planting forests, and so achieve a greener profile. In some contexts, this is a case of greenwashing climate-damaging products, and trading in CO_2 credits can help to prevent necessary change. But in other contexts, these are companies that genuinely want to contribute to reducing the climate crisis and that are working seriously to reduce the climate impact of their own production. One way to increase the competitiveness of companies that do permaculture farming and construction within the existing economic system could be to start selling CO_2 credits. CO_2 credits from permaculture will be of particularly high quality, since in addition to carbon storage, they also regenerate other natural resources.

Large parts of society's greenhouse gas emissions are not regulated by quota systems, including households. Regulating households can be done with a system for personal carbon trading. We can establish such systems at grassroots level, and this is being worked on, for example, in the organization EcoCounts in London.[13, 14] In a carbon trading group, we measure our personal climate footprint. We can thereby learn a lot about how much different actions affect the climate, and it is satisfying to be surrounded by other people who are also doing something active in relation to the climate crisis.

The next step is to set a goal for how much we as a group should emit together and start trading with each other, so that those who emit less than their personal carbon budget can sell their surplus to those who emit more than their personal carbon budget. In this way, we can turn carbon into a currency that works alongside the national currency and values the otherwise externalized climate pollution.

The result will be that goods and services with a low climate footprint will be in demand over those with a high footprint. In connection with such an initiative, we can collaborate with local shops and public services on carbon labeling of goods, and an app can be developed, for example, that makes it easier to register the climate footprint in everyday life.

In addition to making a real difference by reducing the climate footprint, personal carbon trading groups can perhaps make a political difference by raising awareness to the possibility of introducing such a carbon currency at a national and international level, and by creating a body of experience of how it works in practice.

Low-consumption lifestyle and resilience

Within the current monetary system, a low level of consumption will be a realistic foundation for a permaculture lifestyle to be possible with local work and local resources. A life where we achieve resilience through *small scale* with low expenses, low *energy* consumption and the deliberate creation of a high *diversity* of income sources. High expenses tie us to work with high wages from a production system that, to a large extent, contributes to creating climate change and depleting natural resources.

A low-consumption lifestyle compared to the norm in the rich part of the world, on the other hand, helps to limit our consumption of resources and thus create equal distribution.

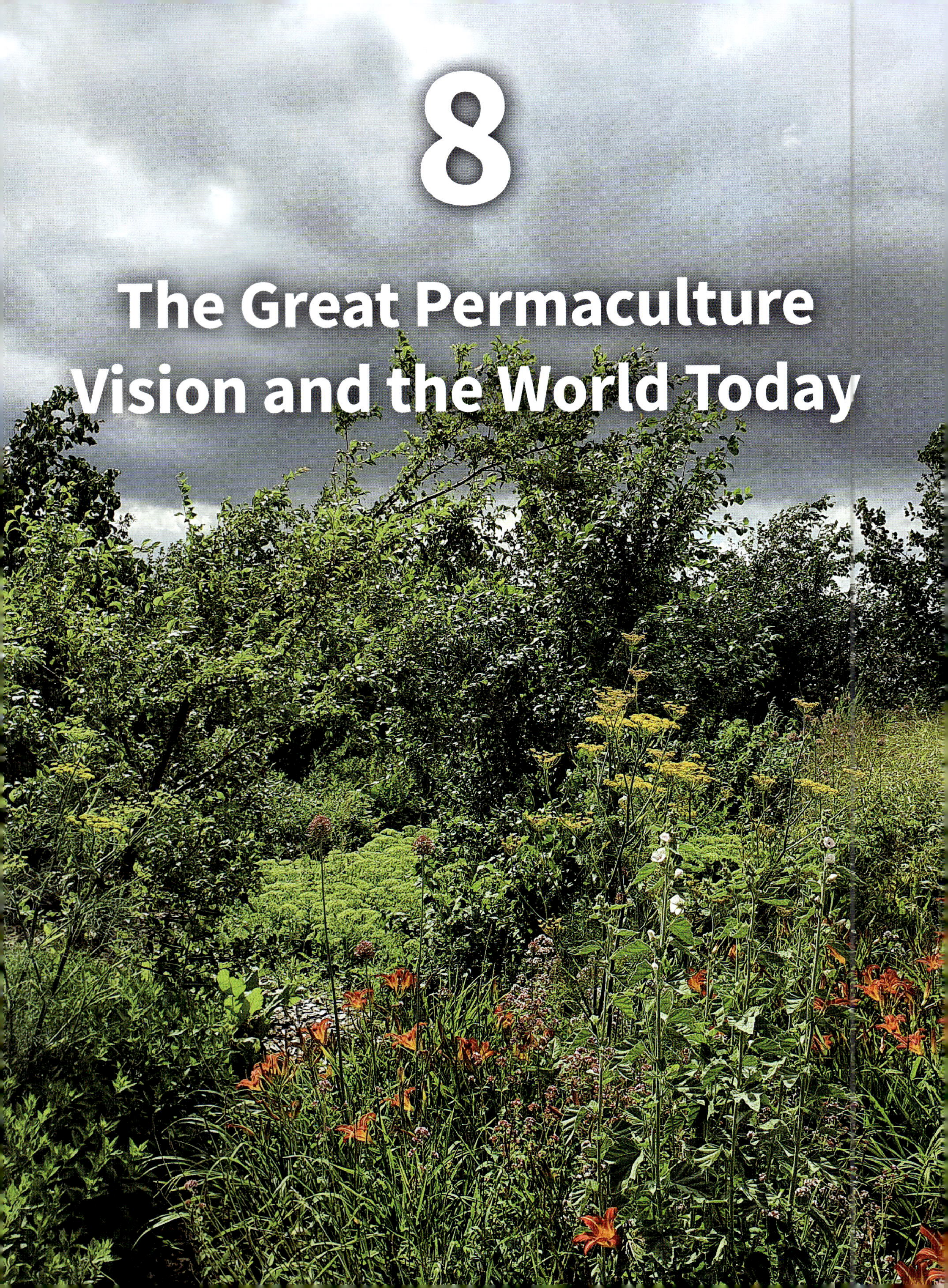

8

The Great Permaculture Vision and the World Today

Summer storm after drought in the food forest. While the grain is flattened in the fields, the sturdy perennial vegetables stand protected in the shelter, the nuts stay on the branches, while a few ripe fruits are torn from the trees. Climate change will bring more extreme weather, and permaculture systems are resilient. Reforest Farm in Skovvirke 2023, Denmark.

All the knowledge gathered in this book shows that the great permaculture vision is possible. It is possible to reforest the Earth, put water back into the landscape, regenerate the resource energy, and improve air quality, and the global community can mitigate climate change. The example of permacultural land use for Denmark can form the basis for permacultural land-use adapted to other countries and continents.

Low-energy houses made of local natural materials are possible as the construction of the future, and the techniques and materials can be used for renovations all over the world.

We can create structures that rebuild local communities, just as we can create monetary systems that promote permaculture and thus regenerate nature and create global equality.

All of this is the result of the development work of many, many enthusiasts. It is clear today that we who are alive now can choose to deal with the crises that global society is drowning in. We can actually choose to transition to a society that regenerates natural resources and counteracts climate change

The period since World War II has been relatively stable with prosperity in many parts of the world, but this has changed, and recent years have been characterized by pandemics, climate disasters, wars and setbacks for democracy and human rights. All of this can be related to humanity's overconsumption of its resource base, but it is not tackled

The land is being reforested. Skovvirke, spring 2016.

politically from this understanding.

Even in a democratic and rich society like Denmark, which aims to be a pioneer in the climate field, it is difficult to get through to politicians with permaculture's holistic solutions on how we mitigate climate change and rebuild natural resources. Mainstream politics insists that the climate crisis must be handled with technological fixes and green growth. While it is difficult to get permaculture through politically in Denmark, we can also see that there are countries in the world where it is even more difficult.

In order to implement the solutions effectively enough, as a rational response to the world situation, world leaders should instead draw up international goals and agreements that commit all countries to permacultural development of local communities.

Many of the initiatives we have proposed in this book may therefore initially seem utopian due to the political circumstances. As permaculturists, however, we must maintain that it is green growth that is an unrealistic utopia and permaculture that is the realistic way to handle the crises and improve the living conditions of the broad masses. It is permaculture that represents security, safety and sustainable welfare.

Permaculture's holistic quality solutions can offer us all hope and opportunities for action, especially when we have moments of fear and discouragement.

It is clear that the transition to permaculture will be in the interest of the vast majority – from rich to poor and to future generations. The vast majority has a common interest in us tackling the five catastrophes. We are the many, and we have the great advantage that we simply have the ethical upper hand. It is an enormous strength that we have science and arguments on our side – and it is only getting clearer and clearer.

Many people are suffering from the current situation, and the danger that we are close to a collapse for climate and biodiversity is real. We can hope that we are close to reaching a historic low point, where it turns around and there is a shift in consciousness, so that we see that we need a global commitment and solidarity to get not only the climate back into balance but also the other planetary boundaries.

If such a shift happens, we are much better equipped if there are permaculture projects all over the world that show how to grow our food, build our houses, provide our energy, etc. Because such experience of what works in practice cannot be built up overnight, it takes time. Although permaculture is excluded from the official education system in most places today, there are many permaculture projects that have experience in dissemination, and it would be a great advantage if permaculture were to be quickly rolled out to the rest of society.

Therefore, with the common awareness of the great permaculture vision, we must constantly work to develop our permaculture projects on all levels, and we must spread them as the engineers of the vision.

We have the knowledge, the will and the strength. We will do our bit to change the course of history and turn thousands of years of work against nature into a positive interaction with nature.

Endnotes

Chapter 1

1 www.oxfamamerica.org/press/press-releases/richest-1-emit-as-much-planet-heating-pollution-as-two-thirds-of-humanity
2 Lin, David, et al.: *Estimating the Date of Earth Overshoot Day*, Global Footprint Network, 2021, p.6.
3 Global Footprint Network, data.footprint-network.org
4 www.un.org/en/climatechange/science/climate-issues/water?
5 en.wikipedia.org/wiki/Groundwater
6 www.eea.europa.eu/publications/europes-groundwater
7 www.who.int/news-room/fact-sheets/detail/ambient-(outdoor)-air-quality-and-health
8 www.bbc.com/future/article/20200504-which-trees-reduce-air-pollution-best
9 Nørretranders, Tor: *Vild verden*, Tiderne Skifter, 2010, p.121.
10 www.fao.org/docrep/u8480e/u8480e0d.htm
11 www.researchgate.net/publication/232673657_Managing_Soils_and_Ecosystems_for_Mitigating_Anthropogenic_Carbon_Emissions_and_Advancing_Global_Food_Security. The figure for land use is 21 years old; the yearly C emissions since then are about 1.36. Now it is 29 billion tons higher, a total of 349 billion tons. The figure for C from fossil fuels is 22 years old, and 8.7 billion tons of C have been added annually, so the figure is now 191 higher, a total of 483 billion tons of C.
12 Whitefield, Patrick: *The Earth Care Manual*, Permanent Publications, 2004, p.43.
13 www.fao.org/global-soil-partnership/resources/highlights/detail/en/c/1458974
14 Sustainable Food Trust: 'Soil degradation a major threat to humanity', 2015, p.2.
15 Simonsen, Hanna Lise: 'Slaget for de vilde bier', *Global økologi*, Summer 2018.
16 www.co2.earth/daily-co2
17 Hansen, James, et al.: 'Target Atmospheric CO_2 Where Should Humanity Aim?', *Open Academic Science Journal* 2(1), 2008.
18 www.ipcc.ch/report/ar6/syr/downloads/report/IPCC_AR6_SYR_SPM.pdf p.20
19 www.ipcc.ch/report/ar6/syr/downloads/report/IPCC_AR6_SYR_SPM.pdf p.12
20 https://climate.copernicus.eu/copernicus-2024-first-year-exceed-15degc-above-pre-industrial-level
21 www.ipcc.ch/report/ar6/syr/downloads/report/IPCC_AR6_SYR_SPM.pdf p.5-7
22 www.climaterealityproject.org/blog/how-feedback-loops-are-making-climate-crisis-worse
23 www.ipcc.ch/report/ar6/syr/downloads/report/IPCC_AR6_SYR_SPM.pdf p.22
24 www.ipcc.ch/2018/10/08/summary-for-policymakers-of-ipcc-special-report-on-global-warming-of-1-5c-approved-by-governments
25 Fog, Kåre: *Økologi – en grundbog*, Gads Forlag, 2001, p.65. Arable land produces on average 6.5 tons dry weight/hectare/year, while a temperate forest produces 10 tons dry weight/hectare/year.
26 Thomsen, Karsten: *Alle tiders urskov*, Nepenthes Forlag, 1996, pp.66–67.
27 Mollison, Bill: *Introduction to Permaculture*, Tagari Publications, 1991.
28 Holmgren, David: *Permaculture Principles & Pathways Beyond Sustainability*, Holmgren Design Service, 2002.
29 Whitefield, Patrick: *The Earth Care Manual*, Permanent Publications, 2004.
30 FAO: 'Stable foods: What do people eat?', www.fao.org/4/u8480e/U8480E07.htm#:~:-text=Just%2015%20crop%20plants%20provide,up%20two%2Dthirds%20of%20this.
31 https://documents1.worldbank.org/curated/en/668871587125098246/pdf/Measurement-Farm-Size-and-Productivity.pdf
32 Rasul, Kajwan, et. al.: 'Energy input and food output: The energy imbalance across regional agrifood systems', *PNAS Nexus* 3(12), 2024.
33 Altieri, Miguel: 'Agroecology Scaling Up for Food Sovereignty and Resiliency', *Sustainable Agriculture Reviews*, vol. 11, 2012, www.researchgate.net/publication/285246565_Agroecology_Scaling_Up_for_Food_Sovereignty_and_Resiliency
34 https://grain.org/en/article/4929-hungry-for-land-small-farmers-feed-the-world-with-less-than-a-quarter-of-all-farmland
35 Whitefield, Patrick: *The Earth Care Manual*, Permanent Publications, 2004, pp.375–388.

Chapter 2

1 https://ourworldindata.org/world-lost-one-third-forests
2 https://extension.psu.edu/how-forests-store-carbon
3 Matthews, Robert, et al.: *Quantifying the sustainable forestry carbon cycle: Summary Report*, Research Agency of the Forestry Commission, 2022, table S2a, p.7.
4 Crawford, Martin: 'A Climate Crisis Solution', *Permaculture*, issue 104, Summer 2020.
5 Project Drawdown: *Farming our way out of the climate crisis*, 2020, fig. 5.2.
6 www.researchgate.net/publication/226203416_Effects_of_windbreaks_on_air-flow_microclimates_and_crop_yields
7 Crawford, Martin: *Creating a Forest Garden*, Green Books, 2010, p.99.
8 Crawford, Martin: *Creating a Forest Garden*, Green Books, 2010, p.91.
9 Crawford, Martin: *How to Grow Perennial Vegetables*, Green Books, 2012, pp.16–17.
10 Toensmeier, Eric, et al.: *Testing the Nutrient Composition of Perennial Vegetables in Denmark, Sweden and the United States*, Perennial Agriculture Institute, 2022.
11 https://ragmans.co.uk
12 https://permakulturgaarden.dk/skovvirke
13 www.oneplanetcouncil.org.uk
14 www.accesstoland.eu
15 https://communitysupportedagriculture.org.uk
16 https://viacampesina.org
17 Heinberg Richard, *Peak Everything*, Clairview 2007, p. 61.
18 Research by Lori Mariono: https://link.springer.com/article/10.1007/s10071-016-1064-4 and a shorter version in Danish: https://videnskab.dk/naturvidenskab/hoens-er-meget-klogere-end-vi-tror.
19 Jensen, Per: *Dyrenes Følelser og vores følelser for dyr*, Forlaget Tro-fast, 2021, p.117.
20 https://journals.plos.org/plosone/article?id=10.1371/journal.pone.0256105
21 www.fao.org/4/y4359e/y4359e0e.htm
22 www.britishkunekunesociety.org.uk/raising-for-meat
23 Nordborg, Maria: *Holistic Management, a critical review of Alan Savory's grazing method*, Sveriges landbruksuniversitet, 2016, p.44.
24 Toensmeier, Eric: *The Carbon Farming Solution*, Chelsea Green, 2016, pp.86–90
25 Nordborg, Maria: *Holistic Management, a critical review of Alan Savory's grazing method*, Sveriges landbruksuniversitet, 2016, p.32.
26 EIP-AGRI Focus Group: *Grazing for Carbon*, European Commission, 2018, p.7.
27 www.ucdavis.edu/food/news/making-cattle-more-sustainable
28 www.landbrugsinfo.dk/public/4/7/1/heste_klimapavirkning
29 www.rewildingeurope.com/wp-content/uploads/2015/07/Natural-grazing---Practices-in-the-rewilding-of-cattle-and-horses.pdf
30 www.landbrugsinfo.dk/-/media/landbrugsinfo/public/6/c/9/guideline_dyreholdere_forvaltning_naturarealer_kvag.pdf
31 Westergaards.dk/kastanje
32 Crawford, Martin: *How to Grow Your Own Nuts: Choosing, Cultivating and Harvesting Nuts in Your Own Garden*, Green Books, 2016, pp.235–237 and 275. For walnuts, we estimate that 50% of the yield is kernel and the rest is husk.
33 https://paulownia.dk/co2-beregning-for-1-hektar-paulownia-traeer
34 Borgen, Anders: 'Anders Borgen forædler flerårige kornsorter til fremtidens permakulturlandbrug', *Tidsskrift om Permakultur*, no. 3, 2010.
35 Toensmeier, Eric: *The Carbon Farming Solution*, Chelsea Green Publishing, 2016, p.149.
36 Nyt fra Danmarks statistik, no. 302, 19.07.2017.
37 5,770,000 inhabitants x 0.0146ha/2435,000ha plowed area x 100 = 3.5%
38 Ibarrola-Rivas, Maria José, et al.: *How Much Time Does a Farmer Spend to Produce My Food?* MDPI, 2016.
39 Smil, Vaclav: *How the World Really Works*, Penguin Books, 2022, p.52.
40 Whitefield, Patrick: *The Earth Care Manual*, Permanent Publications, 2004, pp.268–270
41 Lowth, Henny: *Living Mulches, Final report, 2020–2022*, Organic Research Centre, 2022.
42 Holcomb, Tycho, and Aaen Nolsø, Karoline: 'Småskala Korndyrkning i permanent græseng', *Tidsskrift om permakultur*, no. 24, 2020.
43 Vincent-Caboud, Laura, et al.: 'Using mulch from cover crops to facilitate organic no-till soybean and maize production. A review', *Agronomy for Sustainable Development* 39, 2019.
44 www.wri.org/insights/regenerative-agriculture-good-soil-health-limited-potential-mitigate-climate-change
45 https://drawdown.org/solutions/regenerative-annual-cropping
46 Monbiot, George: *Regenesis, Feeding the World without Devouring the Planet*, Allen Lane, 2022, pp.105–108.
47 www.dirtdoctor.com/organic-research-page/Regenerating-Soils-with-Ramial-Chipped-Wood_vq4462.htm

48 Monbiot, George: *Regenesis, Feeding the World Without Devouring the Planet,* Allen Lane, 2022, pp.109–112.

49 www.fermedubec.com/wp-content/uploads/sites/8/2017/11/ENGLISH-viability_organic_market_garden_without_motorization_paper.pdf

50 Revill, Edward: Articles in *Permaculture,* issue 78, 79 and 80, 2013–2014.

51 Revill, Edward: 'Biokul i skovlandbrug – Biokulsovne samtænkes med allé-dyrkning', *Tidsskrift om Permakultur,* no. 15, 2016.

52 Beck, Martin: *Opbygning af jordens frugtbarhed med kompost,* Økologisk landsforening, 2020.

53 https://rodaleinstitute.org/assets/FSTbooklet.pdf, p.15.

54 Plants For a Future: *Elaeagnus* x *ebbingei* – A plant for all reasons, https://pfaf.org/user/cmspage.aspx?pageid=61.

55 Jacke, Dave, and Eric Toensmeier: *Edible Forest Gardening,* vol. 2, Chelsea Green, 2005, p.535.

56 Annual vegetables in Denmark are given less nitrogen than stated in *Creating a Forest Garden.* Danish norms can be found here: Miljø og fødevareministeriet, Vejledning om harmoni og gødskningsregler, table 1, p.111.

57 www.fao.org/4/X4109E/X4109E03.htm#P277_18693, Chapter 2.4.

58 Luyssaert, S., et al.: 'Old-growth forests as global carbon sinks'. *Nature* 455 (7210), 2008, pp.213–215.

59 www.icos-cp.eu/fluxes/2/forest-carbon-sinks-under-pressure

60 www.icos-cp.eu/fluxes/2/forest-carbon-sinks-under-pressure

61 Díaz-Yáñez, O., et al.: 'Multifunctional comparison of different management strategies in boreal forests', *Forestry: An International Journal of Forest Research,* 93 (1), 2020, pp.84–95.

62 Moberg, Johan: 'Bagsiden af det svenske skoveventyr', *Global Økologi,* Winter 2017.

63 Mudge, Ken, and Steve Gabriel: *Farming the Woods,* Chelsea Green, 2014.

64 *Sådan ligger landet 2022 – Tal om landbruget,* Danmarks Naturfredningsforening, 2022, p.8.

65 Poore, Joseph, and T. Nemecek: 'Reducing food's environmental impacts through producers and consumers', *Science* 360 (6392), 2018.

66 Schou, J.S., et al.: *Hvor mange mennesker kan dansk landbrugs fødevareproduktion brødføde?,* IFRO Udredning, no. 2016/30, 2016.

67 www.agroforestry.co.uk/wp-content/uploads/2021/03/Jeremy_Pasquier_COMPARISON_CALORIC_YIELDS_FOREST_-GARDEN_VS_CONVENTIONAL_AGRICULTURE.pdf

68 Petersen, Jens Kjerulf, et al.: *Vidensyntese om blå biomasse,* DTU Aqua-rapport, no. 387, 2021

69 *Potentiale og planlægning for udtagning af kulstofrige lavbundsjorder med større klimaeffekt,* Ekspertgruppen for udtagning af lavbundsjorder, 2024, p.6

70 *Klimavirkemidler til dansk landbrug,* Seges innovation P/S, 2023. The table on p.6 shows that cattle in Denmark account for an annual climate impact of 4.7 million tons of CO_2e. The reduction is found by: 4.7 x 5/6 = 3.9 million tons of CO_2e. Conversion to a 20-year time perspective: 3.9 x 80 / 28 = 11.1 million tons of CO_2e. It should be noted that this source (https://lf.dk/viden-om/klima/hvad-er-co2) indicates a total emission from Danish cattle of 3.3 million tons of CO_2e, where 1 ton of methane is converted to 25 tons of CO_2e.

71 Crawford, Martin: *How to Grow your Own Nuts – Choosing, cultivating and harvesting nuts in your own garden,* Green Books, 2016, p.235–23 (for walnut we estimate that 50% of the yield is kernels and the rest shells).

72 Crawford, Martin: *How to Grow your Own Nuts – Choosing, cultivating and harvesting nuts in your own garden,* Green Books, 2016, p.275.

Chapter 3

1 Perkins, Richard, *Making Small Farms Work,* 2016, Chapter 4.

2 Weiss, Philipp, and Sjöberg, Annevi: *Skogsträdgården – Odla ätbart överallt,* Hälsingbo Skogsträdgård, 2018, p.42.

3 www.dn.dk/vi-arbejder-for/landbrug/lavbundsjorde

4 Ludwig, Art: *Create an Oasis With Greywater – Choosing, Building and Using Greywater Systems,* Oasis Design, 2009, p34–36.

5 Ludwig, Art: *Create an Oasis With Greywater – Choosing, Building and Using Greywater Systems,* Oasis Design, 2009, pp.65–66.

6 https://kilianwater.com/wp-content/uploads/2020/05/Det-beplantede-filter-og-patogene-bakterier-1.pdf

7 Whitefield, Patrick: *The Earth Care Manual – A permaculture handbook for Britain and other temperate climates,* Permanent Publications, 2004, p.97–98.

8 Illeris, Mira, and Esben Schultz: 'Akvakultur – når fisk, frugt og grønt dyrkes sammen', *Tidsskrift om Permakultur,* no. 13, 2015.

9 Gomes da Silva, Marcelo, et al.: 'Increase of methane emissions linked to net cage fish farms in a tropical reservoir', *Environmental Challenges 5,* 2021.

10 https://theconversation.com/half-of-global-methane-emissions-come-from-aquatic-ecosystems-much-of-this-is-human-made-156960
11 Fengbo, Li, et al.: 'Higher food yields and lower greenhouse gas emissions from aquaculture ponds with high-stalk rice planted', *Environmental Science and Technology* 57 (33), 2023.
12 www.sciencedirect.com/science/article/abs/pii/S0022169422014469
13 https://dm.dk/bio/alle-artikler/klima/blaat-kulstof-fra-de-gemte-og-glemte-skove-rummer-stort-potentiale
14 https://kerteminde.viewer.dkplan.niras.dk/media/3001/blue-carbon-analyse-12-09-2022.pdf
15 www.nature.com/articles/s43017-021-00224-1
16 www.dn.dk/vi-arbejder-for/vand/hav/alegraes
17 https://e-learning.skaldyrcenter.dk/opdraet/blaamuslinger

Chapter 4

1 www.cerulogy.com/wp-content/uploads/2018/02/Cerulogy_Driving-deforestation_Jan2018.pdf, p.5.
2 Holmgren, David, *Permaculture Principles & Pathways Beyond Sustainability*, Holmgren Design Service, 2002, p.xvi.
3 https://richardheinberg.com/museletter-273-neither-utopia-nor-extinction
4 Odum, Howard and Elisabeth: *A Prosperous Way Down*, University Press of Colorado, 2001, p.69.
5 https://ourworldindata.org/energy-production-consumption
6 https://concito.dk/files/media/document/Danmarks%20forbrug%20og%20prioritering%20af%20biomasse%20til%20energiformål.pdf
7 Illeris, Mira, and Esben Schultz: 'Biokulkomfur med skorsten', *Tidsskrift om permakultur*, no. 6, 2011.
8 https://wakefieldbiochar.com/learning-center/how-long-does-biochar-last-in-soil/?srsltid=AfmBOop13Zy5wtlEQE87on4pXAjA9GT-jR472liQFX6CfypwByO_CsaY0
9 https://bezerocarbon.com/insights/the-black-carbon-black-box-introduction-to-black-carbon-and-short-lived-climate-pollutants
10 YouTube: Biochar – Making it in the wood stove and heating our home.
11 Revill, Edward: 'Biochar Stoves', *Permaculture*, issue 78, 2013.
12 Dolleris, Cathrine: 'Fleksovn med røgvasker', *Tidsskrift om Permakultur*, no. 12, 2014.
13 https://exodraft.com
14 www.sciencedirect.com/science/article/pii/S0048969724059965
15 www.aidic.it/cet/14/37/142.pdf
16 www.allpowerlabs.com
17 www.thriveoffgrid.net
18 www.aidic.it/cet/14/37/142.pdf
19 https://klimamonitor.dk/nyheder/art8662371/Solcellers-klimagevinst-er-meget-forskellig-fra-producent-til-producent
20 www.anl.gov/article/solar-panel-manufacturing-is-greener-in-europe-than-china-study-says
21 www.ecowatch.com/solar-environmental-impacts.html
22 www.solacity.com/small-wind-turbine-truth/?srsltid=AfmBOo-qSVzcrz2XR6bJBDuT5L3HaQnhl3UmlB6jE-VgzJK4JYDJ58fNuh
23 https://airbornewindeurope.org
24 https://energypedia.info/wiki/Electricity_Generation_from_Biogas

Chapter 5

1 Illeris, Mira, and Esben Schultz: 'Permakulturhuset – med de mange funktioner', *Tidsskrift om Permakultur*, no. 6, 2011.
2 www.statista.com/statistics/1414973/living-space-per-inhabitant-dwellings-germany
3 www.statista.com/statistics/456925/median-size-of-single-family-home-usa
4 www.kl.dk/analyser/analyser/demografi-og-befolkning/udviklingen-i-den-gennemsnitlige-boligstoerrelse
5 www.researchgate.net/figure/Average-floor-area-per-capita-in-residential-buildings-worldwide-as-of-2008-Data_fig2_358361768
6 https://entranze.enerdata.net
7 www.thelancet.com/action/showPdf?pii=S2542-5196%2824%2900042-1, p.841.
8 Rasmussen, Freja: 'CO_2-udledning fra byggeri - hvad skal der til for at nybyggeri reducerer nok?', *Tidsskrift om Permakultur*, no. 25, 2021
9 Marsh, Rob, et al.: *Arkitektur og miljø*, Arkitektskolens forlag, 2000, p.21.
10 Sommer, Morten O.A.: *Er mit liv bæredygtigt? Sådan påvirker dine handlinger miljøet*, Strandberg Publishing, 2024, pp.58–59.
11 Rempel, Alexandra R., et al.: Supplementary information: Magnitude and distribution of the untapped solar space-heating resource at climatic and metropolitan scales, pp.9–11 (an appendix to www.opb.org/article/2021/11/30/oregon-research-skylights-energy-source).

12 Rempel, Alexandra R., et al.: 'Interpretation of passive solar field data with EnergyPlus models: Un-conventional wisdom from four sunspaces in Eugene, Oregon', *Building and Environment* 60, pp.158–172.
13 www.opb.org/article/2021/11/30/oregon-research-skylights-energy-source
14 Ideas for different types of movable insulation can be found in this book: Langdon, William K.: *Movable Insulation: A Guide to Reducing Heating and Cooling Losses Through the Windows in Your Home,* Rodale Press, 1980.
15 https://cchrc.org/wp-content/uploads/2019/10/window_insulation_final.pdf
16 Rempel, Alexandra, et al.: 'Improving the passive survivability of residential buildings during extreme heat events in the Pacific Northwest', *Applied Energy* 321, 2022, 119323.
17 www.sciencedirect.com/science/article/abs/pii/S1364032121008753?via%3Dihub#!
18 Abrahamsson, Karen: 'Vandtoiletter og rensningsanlæg ... er noget forfærdeligt svineri', T*idsskrift om Permakultur,* no. 13, 2015.
19 www.ecosanres.org/pdf_files/ESR2010-1-PracticalGuidanceOnTheUseOfUrineInCropProduction.pdf
20 https://sswm.info/factsheet/urine-fertilisation-%28small-scale%29
21 Apple and pear trees can be fertilized with 140kg N/ha and hazel with 85kg N/ha. For nutrient norms see: https://lbst.dk/Media/638612176728198485/RETTET%20-%20Vejledning_om_goedskning_og_harmoniregler_2024_2025.pdf, p.135.
22 The feces of a person contain 0.73kg P/year. Source: Harper and Halestrap: Lifting the Lid, Centre for Alternative Technology Publications, 1999, p.122.
23 The phosphorus norm for common alder is 15kg/ha/year. Source: https://lbst.dk/Media/638612176728198485/RETTET%20-%20Vejledning_om_goedskning_og_harmoniregler_2024_2025.pdf, p.136.
24 Rasmussen, Freja: 'CO_2 udledning fra byggeri – hvad skal der til for at nybyggeri reducerer nok?', *Tidsskrift om Permakultur,* no. 25, 2021.
25 www.weforum.org/stories/2024/09/cement-production-sustainable-concrete-co2-emissions
26 Science for environment policy issue 38: The significance of embodied carbon and energy in house construction.
27 Pöyry, Amalia, et al.: 'Embodied and construction phase greenhouse gas emissions of a low-energy residential building', *Procedia Economics and Finance* 21, 2015. www.sciencedirect.com/science/article/pii/S2212567115001872
28 Rasmussen, Freja: 'CO_2 udledning fra byggeri – hvad skal der til for at nybyggeri reducerer nok?', *Tidsskrift om Permakultur,* no. 25, 2021.
29 For the calculations, data on the climate footprint of various building materials was obtained from www.oekobaudat.de/en.html.
30 Illeris, Mira, and Esben Schultz: 'Permakulturhuset – med de mange funktioner', *Tidsskrift om Permakultur,* no. 6, 2011.
31 Racusin, Jacob Deva, and Ace McArleton: *The Natural Building Companion,* Chelsea Green Publishing, 2012, pp.225–232.
32 www.biochar-journal.org/en/ct/3
33 Jones, Barbara: *Building with Straw Bales,* Green Books, revised and updated edition, 2009.

Chapter 6

1 Mollison, Bill: *Permaculture – A Designers' Manual,* 2nd edition, Tagari Publications 1988, p.510.
2 Sommer, Morten O.A.: *Er mit liv bæredygtigt? Sådan påvirker dine handlinger miljøet,* Strandberg Publishing, 2024, pp.36–37.
3 www.britishapplesandpears.co.uk/wp-content/uploads/2022/07/Li_et_al-2022-Nature_Food.pdf
4 https://transitionnetwork.org
5 Hervé-Gruyer, Perine, and Charles Hervé-Gruyer: *Miraculous Abundance, One Quarter Acre, Two French Farmers, and Enough Food to Feed the World,* Chelsea Green Publishing, 2016, Chapter 11.
6 Morgan, Faith. *The Power of Community: How Cuba Survived Peak Oil.* Yellow Springs, OH: Arthur Morgan Institute for Community Solutions, 2006. 53 min. https://youtu.be/keMDUUHgXbo.
7 www.landcoalition.org/en/uneven-ground/executive-summary
8 https://openknowledge.fao.org/server/api/core/bitstreams/78a9a62d-a96d-4bc9-be4e-e509e2c51693/content, p.7.
9 Unpicking food prices: Where does your food pound go, and why do farmers get so little?, www.sustainweb.org, 2022.
10 https://sc-fss2021.org/wp-content/uploads/2021/06/UNFSS_true_cost_of_food.pdf
11 https://sc-fss2021.org/wp-content/uploads/2021/06/UNFSS_true_cost_of_food.pdf, p.21.
12 https://openknowledge.worldbank.org/entities/publication/7ee-f73a7-c4f6-590b-8e31-3a4051d875a9
13 www.accesstoland.eu/Overview-of-Community-Supported-Agriculture-in-Europe
14 Christian, Diana Leafe: *Creating a Life Together,* New Society Publishers, 2003.

15 Landemore, Hélène: *Open Democracy – Reinventing Popular Rule for the Twenty-First Century*, Princeton University Press, 2020, pp.61–66 and 76–77.
16 https://da.wikipedia.org/wiki/Sociokrati
17 Rau, Ted, *Sociocraty – A Brief Introduction.* Sociocraty For All, 2023.

Chapter 7

1 Jackson, Andrew and Ryan-Collins, Josh, et al.: 'Where Does Money Come From?: A Guide to the UK Monetary & Banking System', New Economics Foundation, 2014, and https://positivemoney.org.
2 https://positivemoney.org
3 Ryan-Collins, Josh, et al.: 'Where Does Money Come From?: A Guide to the UK Monetary & Banking System' New Economics Foundation, 2014, and https://positivemoney.org/videos/inequality-why-are-the-rich-getting-richer.
4 https://positivemoney.org/videos/inequality-why-are-the-rich-getting-richer
5 Rersgaard Nielsen, Tune: Den stigende ulighed drives frem af pengesystemet, www.jak.dk/stigende-ulighed-drives-frem-pengesystemet.
6 https://themoneycharity.org.uk/money-statistics-archive
7 Lietaer, Bernard: *Fremtidens penge*, Borgen, 2002, p.380.
8 Drawing from video: https://positivemoney.org/videos/inequality-why-are-the-rich-getting-richer.
9 Losmann, Carmen: *Oeconomia*, Germany DCP, 2020 (English subtitles) and Barmes, David (lead), and Fran Boait, 'The Tragedy of Growth', PositiveMoney, 2020
10 Nielsen, Jørgen Steen: *Den store omstilling*, Informations forlag, 2013.
11 https://positivemoney.org
12 North, Peter: *Local Money*, Transition Books, 2010, Chapter 5.
13 https://ecocore.org/agenda-for-change
14 https://ecocounts.community/about

Index

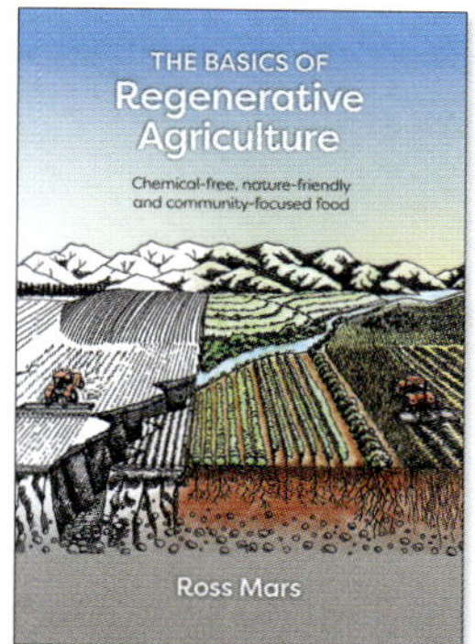

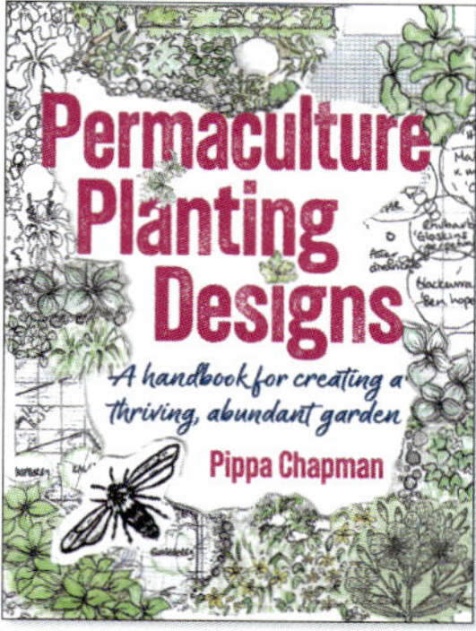

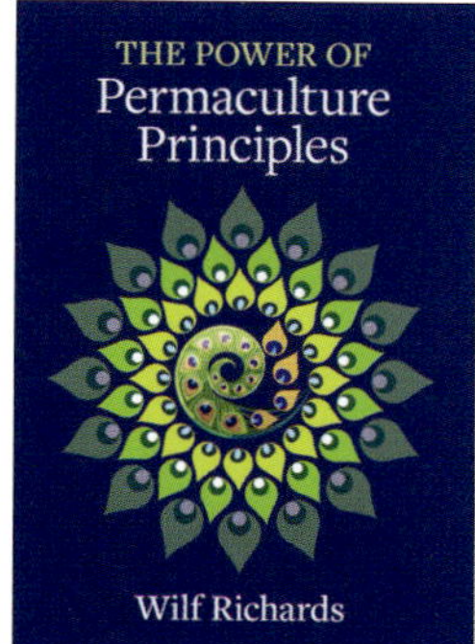

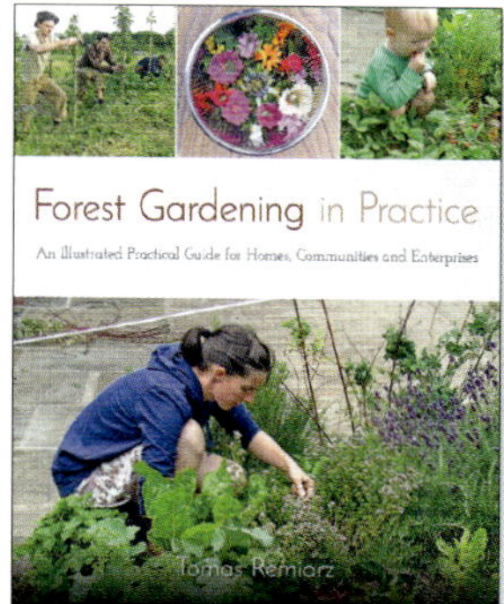

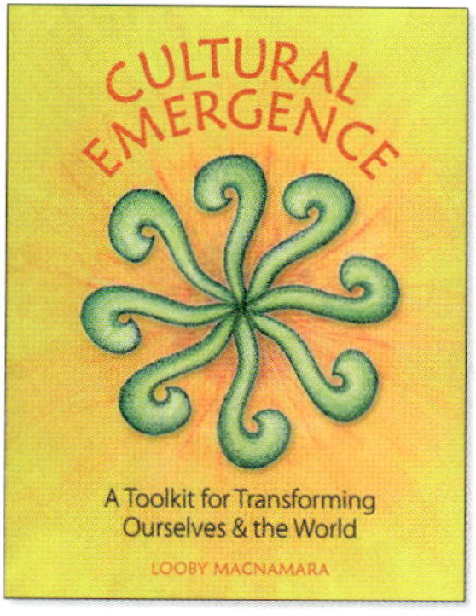